Earthquake design practice for buildings
second edition

Edmund Booth
MA, CEng, FICE, FIStructE

and

David Key
Phd, CEng, FICE, FIStructE

Published by Thomas Telford Publishing, Thomas Telford Ltd, 1 Heron Quay, London E14 4JD.
www.thomastelford.com

Distributors for Thomas Telford books are
USA: ASCE Press, 1801 Alexander Bell Drive, Reston, VA 20191-4400, USA
Japan: Maruzen Co. Ltd, Book Department, 3–10 Nihonbashi 2-chome, Chuo-ku, Tokyo 103, Japan
Australia: DA Books and Journals, 648 Whitehorse Road, Mitcham 3132, Victoria, Australia

First edition 1988
Second edition 2006
Reprinted 2008

A catalogue record for this book is available from the British Library

ISBN: 978 0 7277 2947 7

Typeset by Academic + Technical, Bristol
Printed and bound in Great Britain by MPG Books, Bodmin

Contents

Sources for photographs

(Note: where no source is indicated, the source is the authors.)

Fig. 1.1   Port and Airport Research Institute, Japan.

Fig. 1.2   Colin Taylor, Department of Civil Engineering, Bristol University.

Fig. 1.3   Edmund Booth.

Fig. 1.4   Earthquake Engineering Field Investigation Team (EEFIT), UK.

Fig. 1.6   Mike Winney.

Fig. 1.7   Edmund Booth.

Fig. 1.8   Edmund Booth.

Fig. 1.10   Antonios Pomonis (Cambridge University).

Fig. 1.13   J. Meehan. Courtesy of Karl V. Steinbrugge Collection, Earthquake Engineering Research Center, University of California, Berkeley.

Fig. 1.14   Robin Spence.

Fig. 1.16   EEFIT UK.

Fig. 1.18   Peter Yanev, courtesy of ABS Consulting.

Fig. 1.24   Peter Merriman, BNFL Ltd.

Fig. 2.9   James Jackson, University of Cambridge.

Fig. 4.6   Karl V. Steinbrugge Collection, Earthquake Engineering Research Center, University of California, Berkeley.

Fig. 5.3   David G. E. Smith.

Fig. 7.2   Jack Pappin.

Fig. 8.5   R. C. Fenwick.

Fig. 8.22   Peter Yanev, courtesy of ABS Consulting.

Fig. 9.2   Peter Yanev, courtesy of ABS Consulting.

Fig. 9.7   Courtesy of the SAC Project 7.03 – Georgia Tech (R. Leon and J. Swanson).

Fig. 10.2   Antonios Pomonis (Cambridge University)

Fig. 10.5   D. D'Ayala.

Figs 13.3, 13.4, 13.8   Courtesy of ALGA SpA.

Fig. 13.5   (a) Courtesy of Arup (b) courtesy of Frank la Riviere.

Fig. 13.6   L. Megget.

Fig. 14.3   C. Perry, E. Fierro, H. Sederat and R. Scholl (1993). Seismic upgrade in San Francisco using energy dissipation devices. *Earthquake Spectra*, Vol. 9, No. 3, pp. 559–579. Courtesy of EERI.

Preface

Scope of the book

This book is intended as a design guide for practitioners and advanced students with a sound knowledge of structural design who are not expert in seismic aspects of design, and perhaps are encountering the problem for the first time. Earthquake engineering is a vast subject and the intention of this book is not to provide a fully comprehensive treatment of all its aspects. Rather, it is to provide the practising engineer with an understanding of those aspects of the subject that are important when designing buildings in earthquake country, with references to sources of more detailed information where necessary. Many of the principles discussed also apply to the design of non-building structures, such as bridges or telecommunications towers, but the scope of this book is restricted to buildings.

Although earthquakes do not respect national boundaries, the practice of earthquake engineering does vary significantly between regions, and this is reflected in the differing formats and requirements of national seismic codes. The book is intended to be more general than to describe the approach in just one code, although it reflects the experience of the authors, particularly of the European seismic code Eurocode 8 and of US codes. Japanese practice is in many ways very different, and is scarcely mentioned here.

Outline

Earthquakes regularly occur which test buildings much more severely than their designers might reasonably have expected, and earthquake engineers should (and do) make use of this chance (found much more rarely in other disciplines) to find out whether the current theories actually work out in practice. The first chapter therefore reviews the lessons from earthquake damage for designers of buildings. Chapter 2 is a brief introduction to engineering seismology, including such matters as measuring earthquakes and the ground motions they produce. Chapter 3 outlines the important principles of structural dynamics applicable to seismic analysis, and Chapter 4 discusses the analysis of soils (a crucial issue where the soil provides the dual and conflicting roles of both supporting and also exciting the structures founded on it). Chapter 5 presents the fundamentally important issue of the conceptual design of buildings; if this is wrong, it is unlikely that the seismic resistance will be satisfactory. Chapter 6 gives an introduction to some seismic codes of practice. Chapter 7 discusses the design of foundations, while Chapters 8 to 11 discuss issues specific to seismic design in the four main materials used for building structures – concrete, steel, masonry and timber. So far, the book has concentrated on the primary structure of a building, but its contents are also important and can suffer as much or even greater damage in an earthquake. Chapter 12 therefore discusses building contents and cladding. Chapter 13 introduces special measures to improve earthquake resistance, such as mounting buildings on base isolation bearings or introducing various types of devices to increase structural damping. Existing buildings without adequate seismic resistance pose a huge safety and economic threat in many parts of the world and the final chapter discusses how to assess and strengthen them.

Acknowledgements

Assistance in preparing the text and illustrations is gratefully acknowledged from many friends and colleagues. Particular thanks are due to Richard Fenwick for permission to base parts of Chapter 8 on material originally prepared by him, and to Jack Pappin in a similar way for material used in Chapters 4 and 7. Richard and Jack also provided many helpful and detailed comments on the text, as did Dina D'Ayala, Ahmed Elghazouli, James Jackson, David Mallard, Agostino Marioni, Alain Pecker, Bryan Skipp, Robin Spence and David Trujillo.

Introduction to the first edition

This book deals with earthquakes, which are natural disasters. In a letter to *The Times*, on 13 July 1984, the Archbishop of York wrote

> 'Disasters may indeed be messengers, in that they force us to think about our priorities. They drive us back to God. They remind us of mistakes and failures, and they call forth reserves of energy and commitment which might otherwise remain untapped. Disasters also remind us of the fragility of life and of our human achievements.'

Designing for earthquake resistance is difficult, not because the basic steps in the process are necessarily hard, but because the fundamental concept of earthquake resistance is different from design for other loadings, such as wind pressure or gravity loads. It is different in two important respects. Firstly, it is a dynamic loading involving a number of cyclic reversals, so that the behaviour of the structure involves an understanding of structural dynamics. Secondly, normal design practice accepts that, in response to a major earthquake, a building structure may suffer major damage (but should not collapse), whereas for wind and gravity loads even minor damage is not acceptable.

Earthquake-resistant design is not widely taught. For the practising engineer it is a difficult subject to come to grips with, not because there is a shortage of information, but because there is a surfeit. It is a subject where it is possible to drown in information and to starve for knowledge. Professor G. Housner, in an address to the participants at the Eighth World Conference on Earthquake Engineering in 1984, suggested that, if the current logarithmic increase in the number of papers presented at the four-yearly World Conferences continued, by the 19th it would take four years to present the papers.

The author himself (David Key) has struggled over many years to develop a sound approach to the design of structures in earthquake zones. This book is intended to guide others not only in the basic procedures of design but also to point out sources of specialised information on the subject when it is beyond the scope of this work.

Earthquake engineering has to a large extent slipped out of the hands of the practical designer, and into the hands of the specialist, who usually employs a suite of computer programs to provide great quantities of unnecessarily precise information on such subjects as the ground motion spectrum or the dynamic response of the building to some long past earthquake which can only bear the vaguest resemblance to any ground motion to which the building could be subjected. In the author's view the principal ingredients in an earthquake-resistant design can be categorised as follows.

Essential

(*a*) a sound structural concept

(*b*) an understanding of the way in which the structure will behave when primary structural elements have yielded

(*c*) an approximate idea of the peak ground acceleration likely to be experienced, and the predominant frequency

(*d*) the application of engineering common sense to the fact that the building may be violently shaken

(*e*) good detailing

(*f*) good quality construction and inspection.

Useful

(*a*) detailed elastic analysis of the structure

(*b*) dynamic analysis of simple models

(*c*) a soil–structure interaction study when justified by the soil and structure properties

(*d*) estimates of the ground motion spectrum.

The designer is in the end the person who puts all the theory into steel and concrete, and who bears the responsibility for it.

This book assumes a competent knowledge of structural design by the reader. It is intended as a guide to the normal processes of design, and to provide directions for further study when the structural problem is out of the ordinary.

David Key, 1988

Introduction to the second edition

Many things have changed since David Key wrote his introduction to the first edition in 1988, but his approach as outlined above remains just as valid. The major changes in seismic engineering can be listed as follows.

(1) Publication of a European seismic code of practice and significant developments in codes elsewhere, including the USA.

(2) A vast increase in the number, availability and quality of earthquake ground motion recordings, and a better understanding of the influence of soils and earthquake characteristics on ground motion.

(3) A greater appreciation of the factors that need to be accounted for in the seismic design of steel structures.

(4) Transformation of non-linear time-history analysis from a specialist research method to a potentially useful (and actually used) tool for practising engineers.

(5) Development of non-linear static (pushover) techniques of analysis.

(6) Development of practical methods for assessing and improving the seismic resistance of existing structures.

(7) Much greater use and experience of seismically isolated structures and those with added structural damping, although they still represent only a tiny minority of structures actually built.

(8) Improved ability to predict the response of soils to earthquake loading, including their potential for liquefaction.

The second edition has therefore retained the same basic structure and intention of the original edition, but all sections have been partially or (in most cases) wholly rewritten to reflect the changes noted above. The scope has been limited to buildings, so the chapter in the first edition covering bridges, tanks, towers and pipelines has been removed, and replaced with one on the assessment and strengthening of existing buildings.

Edmund Booth, 2005

Foreword

In the introduction to the first edition of *Earthquake design practice for buildings*, David Key memorably wrote

> 'Earthquake engineering has to a large extent slipped out of the hands of the practical designer, and into the hands of the specialist, who usually employs a suite of computer programs to provide great quantities of unnecessarily precise information . . .'

and it was partly for this reason that he directed that first edition to the needs of the practical designer, not to those of the earthquake specialist.

In the intervening 17 years the science of earthquake engineering has advanced enormously, and today it is inconceivable that a large building project would be built in an earthquake area without the advice of a specialist. Indeed Edmund Booth who, with David Key, has so admirably expanded and updated this book, is one of today's leading earthquake engineering specialists. But the resulting book is not written for the specialist. It is remarkable in the way it adheres to the main goal which motivated David Key in the first place – to make earthquake engineering intelligible and interesting to the non-specialist, practical designer.

Today there is of course much more ground to cover than there was in 1988 – the development of codes, the improved understanding of ground motion, new methods of analysis and many innovations in providing for earthquake resistance – and these are all succinctly covered in this new edition with admirable clarity.

But the key features that made the first edition so valuable are still present. First, that the approach to earthquake engineering presented derives from the authors' direct observation of the damage to buildings in large earthquakes; the principal modes of damage are clearly identified, and many very well chosen photographs are used to illustrate these. This experience is used to inform the design guidance given.

Second, the book does not depend on a heavily mathematical approach. Rather, equations are used sparingly and the authors rely on good, clear descriptions of structural behaviour, backed by excellent diagrams, making the text accessible to all those who have to deal with the design of buildings structures for earthquake areas, whether as engineers or architects.

Third, the book is based on long personal experience by both authors of the design of buildings in earthquake areas worldwide, and can thus give authoritative advice on the appropriate codes, design procedures and structural arrangements to adopt for both highly seismic areas and areas of low seismicity. This is advice we can rely on.

Special features of this edition which will make it particularly valuable to engineering designers are:

- its timely account of the Eurocodes, now finally becoming published documents and soon to become mandatory in some areas, with which Edmund Booth has been closely involved
- the excellent chapter on conceptual design, setting out some fundamentals which should be thought about while a building's form and siting are still being developed, and which architects as well as engineers will find illuminating
- a valuable new chapter on the assessment and strengthening of existing buildings, an activity whose importance is already growing in many countries, as we look for ways to protect our urban centres from future earthquake disasters
- an excellent state of the art on seismic isolation, rightly identified by the authors as 'an idea whose time has come'.

However, as well as being a practical guide to design, the book is also a valuable reference work, offering excellent bibliographies on all the major topics, and valuable suggestions for follow-up study where needed.

For these reasons and many more this book will be appreciated – and enjoyed – by all those who have responsibility for the design, construction and maintenance of buildings in earthquake areas, both in the European area and worldwide.

Professor Robin Spence
President, European Association for Earthquake Engineering
Cambridge
July 2005

Notation

Notes

(1) The units shown for the parameters are to indicate the dimensions of the parameters, but other consistent systems of units (involving for example the use of millimetres instead of metres) would also be possible.

(2) Notation not given in this table is defined at the point of occurrence in the text.

Symbol	Description
a_g	Peak ground acceleration: m/s^2
b	Width of compression flange of concrete beam: m
b_f	Breadth of flange of steel section: m
c_u	Undrained shear strength of soil: kN/m^2; Dimensionless coefficient in the US code ASCE 7 relating to the upper limit on calculated period of a building
d	Effective depth to main reinforcement in a concrete beam: m; Diameter of bolt or other fastener joining timber members: m
d_b	Diameter of reinforcing steel in concrete: m
d_r	Relative displacement between points of attachment of an extended non-structural element: m
e	Length of the shear link in an eccentrically braced frame (EBF): m
F	Force: kN
f'_c	Cylinder strength of concrete: kN/m^2
f'_{cc}	Compressive strength of concrete under confining pressure f_1: kN/m^2
f_1	Hydrostatic confining pressure on an element of concrete: kN/m^2
F_a	Horizontal force on non-structural element: kN
F_b	Seismic shear at base of building: kN
f_b	Compressive strength of masonry: kN/m^2
$F_{elastic}$	Seismic force developing in an elastic (unyielding) system: kN
F_i	Force at level i: kN
$F_{plastic}$	Seismic force developing in a plastic (yielding) system: kN
F_y	Yield force: kN
f_y	Yield strength of steel: kN/m^2
g	Acceleration due to gravity n/s^2
G_0	Shear modulus of soil at small strains: kN/m^2
G_s	Shear modulus of soil at large shear strain: kN/m^2
H	Building height: m
h	Minimum cross-sectional dimension of beam: m; Greater clear height of an opening in a masonry wall: m

h_{ef}	Effective height of a masonry wall: m
h_s	Clear storey height of shear wall between lateral restraints: m
h_w	Overall height of shear wall: m;
	cross-sectional depth of beam: m
k	Spring stiffness: kN/m;
	Dimensionless exponent in equation 6.2 for distribution of seismic forces with height;
	Dimensionless empirical constant in Table 10.5
K_{eff}	Secaut stiffness of a non-linear system at a given deflection: kN/m (see Figure 3.24)
L	Length of a masonry wall: m
l	Effective unrestrained length of a beam or column: m
L^*	Critical span of beam corresponding to formation of plastic hinges within span under lateral loading: m
L'	Clear span of beam: m
l_{av}	Average length of shear walls in a building: m (see Table 10.5)
L_i	Structural property defined in equation 3.11: tonnes
L_{pl}	Effective plastic hinge length: m
L_v	Bending moment to shear force ratio at the critical section of a plastic hinge forming in a concrete member
M	Magnitude of earthquake;
	Mass: tonnes
M_s	Magnitude of earthquake measured using the surface wave scale
$m(x)$	Mass per unit length at height x: kN/m
$\boldsymbol{M}_A, \boldsymbol{M}_B$	Plastic hinge moments forming at either end of a beam: kNm
M_i	Structural property defined in equation 3.12: tonnes
m_i	Mass at level i: tonnes
$\boldsymbol{M}_p$	Flexural strength of the shear link in an eccentrically braced frame (EBF): kN-m
M_u	Bending moment in a plastic hinge under ultimate conditions: kNm
N	Blow count per 300 mm in the Standard Penetration Test (SPT)
N_1 (60)	Corrected SPT blow count: see section 4.3.2(d–f)
n	Number of storeys in a building
$\boldsymbol{P}$	Axial load in a column: kN
P_1	Probability of exceedence in one year
P_y	Probability of exceedence in y years
q	'Behaviour' or force reduction factor for structural systems in Eurocode 8
q_a	'Behaviour' or force reduction factor for non-structural elements in Eurocode 8
R	'Response modification' or force reduction factor for structural systems in the US code IBC;
	Radius of a friction pendulum isolation bearing: m
r_y	Radius of gyration of a beam or column about its minor axis: m
S	Soil amplification factor in Eurocode 8
S_a	Spectral acceleration: m/s^2

S_{ai}	Spectral acceleration corresponding to the period of mode i: m/s^2
S_d	Spectral displacement: m
$S_e(T)$	Spectral acceleration, based on elastic response, corresponding to structural period T: m/s^2
S_v	Spectral velocity: m/s
T	Return period: years;
	Structural period: s
T_1, T_2, T_3	Periods of first, second, third modes of building: s
T_a	Fundamental vibration period of non-structural element: s;
	Empirically determined vibration period of a building: s
T_B, T_C	Periods defining the peak of the design response spectrum in Eurocode 8: s
t_{ef}	Thickness of a masonry wall: m
T_{eff}	Effective period of a non-linear system at a given displacement: s
t_f	Thickness of flange of steel section: m
$u_{elastic}$	Seismic displacement of elastic (unyielding) system: m
$u_{plastic}$	Seismic displacement of a plastic (yielding) system: m
u_{ult}	Displacement at ultimate capacity: m
u_y	Displacement at yield: m
v	Masonry shear strength under zero compressive load: kN/m^2
V_1, V_2, V_3	Seismic shears at base of building corresponding to first, second, third modes: kN
v_d	Design in-plane shear strength of masonry: kN/m^2
V_p	Shear capacity of the shear link in an eccentrically braced frame (EBF): kN
V_u	Shear force in a plastic hinge under ultimate conditions: kN
W_a	Weight of non-structural element: kN
X	Dimensionless reduction factor
x	Height above fixed base: m
z	Total height of building above base: m
z_i	Height above base of level i: m
α_{sl}	Dimensionless empirical constant in equation 8.4 for plastic hinge length
δ	Lateral deflection: m
$\phi_i(x)$	Modal deflection at height x in mode i
ϕ_p	Curvature of a plastic hinge at rotation θ_p: radians/m
ϕ_u	Ultimate curvature of a plastic hinge: radians/m
ϕ_y	Curvature of a plastic hinge at first yield: radians/m
γ	Shear strain
γ_a	Importance factor for non-structural element, in Eurocode 8
γ_m	Partial factor on material strength
η	Correction factor to adjust response for damping other than 5%
μ	Displacement ductility;
	Coefficient of friction
ν	Reduction factor in Eurocode 8 to convert design displacements at ultimate limit state to serviceability limit state

θ_p — Plastic rotation of a plastic hinge: radians

θ_u — Ultimate rotation at a plastic hinge: radians

θ_y — Rotation at a plastic hinge at yield: radians

ρ — Ratio of tension reinforcing steel area to cross-sectional area of concrete member;
Ratio of force demand on an element to capacity of the element

ρ' — Ratio of compression reinforcing steel area to cross-sectional area of concrete member

σ_v — Vertical stress in masonry due to permanent loads: kN/m^2

σ_{vo} — Total vertical stress in soil at the level of interest due to gravity loads: kN/m^2

σ'_{vo} — Effective vertical stress in soil at the level of interest due to gravity loads: kN/m^2

τ_e — Effective shear stress in soil under design earthquake loading: kN/m^2

ξ — Percentage of critical damping

Ω — Minimum ratio of resistance moment to design moment at plastic hinge position

1 The lessons from earthquake damage

'The bookful blockhead, ignorantly read,
With loads of learned lumber in his head.'

An essay on criticism, Alexander Pope

1.1 Damage studies

The study of earthquake damage was the original source of design criteria for earthquake-resistant structures. For example, following the 1906 San Francisco earthquake, engineers observed that buildings designed to withstand a wind force of $30 \, \text{lbf/ft}^2$ performed well. That simple observation embodied a great deal of common sense, including the concept of using a static lateral force to reproduce the effect of an earthquake.

The reason for the quotation at the start of this chapter is to emphasise the view that earthquake engineering is not to be learned from books only. Engineers generally have some experience of their structures being loaded to something approaching their design load, and errors in calculation or implementation will show up in the form of major or minor defects. In this way there is some feedback from experience and some encouragement to use this experience. For earthquake design this is rarely the case, so that the only source of experience for engineers is either the study of damage reports or, even better, in carrying out a damage survey themselves. To take the subject of earthquake-resistant design out of the realms of a book-taught subject into the realms of thoughtful engineering it is essential that as much practical knowledge as possible is included. Designers need to feel what may happen to their structures as well as to know a set of design rules.

Engineers are most accustomed to static loads. One of the most important lessons learned from damage surveys is the difference in failure patterns between static loads applied in a single direction and those due to cyclic loading. There are important differences in the way that failure modes develop between the two.

An important aspect of post-earthquake study is the realisation of the important role that the quality of construction plays. Earthquakes are no respecters of theories, calculations or divisions of responsibility. Many instances of poor-quality construction are invariably exposed in damaged buildings. Badly placed reinforcement, poorly compacted concrete and incomplete grouting of masonry are some of the commonest examples.

The immediate human response to earthquakes is not in general regarded as a design criterion. Nevertheless, every earthquake shows up numerous examples of lives at risk from minor faults in construction – falling masonry or cladding,

Table 1.1 Long-term human response to earthquakes

Stage	Time	Event	Reaction	
			Positive	Negative
1	0–1 minute	Major earthquake		Panic
2	1 minute–1 week	Aftershocks	Rescue and survival	Fear
3	1 week–1 month	Further aftershocks	Short-term repairs	Allocation of blame – builders, designers, officials etc.
4	1 month–1 year		Long-term repairs and pressure for higher standards	
5	1 year–10 years			Diminishing interest
6	10 years to the next time			Reluctance to meet costs of seismic provisions, research etc. Increasing non-compliance with regulations
7	The next time	Major earthquake	Repeat stages 1–7	

ceiling tiles dislodged, window frames separating from the walls and toppling inwards or outwards, and escape paths blocked by jammed doors and fallen masonry.

In the longer term, human response follows the pattern shown in Table 1.1, and while this might be seen as light-hearted or cynical there is no doubt that, as the time of the last disaster recedes, it becomes increasingly difficult to convince owners, officials and professionals of the need for earthquake-resistant measures. The task of the building design team is not always neatly prescribed by sets of regulations, and the achievement of high technical standards requires a clear understanding of the problem and mutual support in presenting it to owners and officials.

1.2 Ground behaviour

The effects of violent shaking on the ground are temporarily to increase lateral and vertical forces, to disturb the intergranular stability of non-cohesive soils and to impose strains directly on surface material locally if the fault plane reaches the surface.

The results of a transient increase in lateral and vertical forces means that any soil structures that are capable of movement are at risk. The resulting types of damage are landslips and avalanches, and experience of the 1970 earthquake in Peru and the 1964 earthquake in Anchorage, Alaska shows that these may be

Fig. 1.1 Damage in the port of Kobe, Japan, 1995

on a massive scale. One village, Yungay, in Peru was destroyed almost entirely with
the loss of 18 000 lives by a debris flow involving tens of millions of tons of rock
and ice.

The disturbance of the granular structure of soils by shaking leads to consolidation of both dry and saturated material, due to the closer packing of
grains. For loose saturated sands the pore pressure may be increased by shaking
to the point where it exceeds the confining soil pressure, resulting in temporary
liquefaction. This is an important effect as it can lead to massive foundation failure
in bearing and piled foundations, the collapse of slopes, embankments and
dams, and to the phenomenon of 'boiling' where liquefied sand flows upwards in
surface pockets. Dockside structures are found to be particularly susceptible to
liquefaction-induced failure, because loose saturated granular soils are often
present as fill or foundation materials. Liquefaction-induced failure caused
many billions of pounds of direct damage to the Port of Kobe Japan in the 1995
earthquake, with a similar amount in lost revenue due to closure of the port
(Fig. 1.1).

Shear movements in the ground may be at the surface or entirely below it. If
the earthquake fault reaches the surface, permanent movements of considerable
magnitude occur, which can amount to several metres in large earthquakes
(Fig. 1.2). Surface shear movements may also take place as a result of other
causes – landslips or consolidation for example. Subsurface shear failures can
occur in weaker strata, leading to damage of embedded or buried structures.
Subsurface shear failures can also reduce the transmission of ground motion to
the surface, effectively putting an upper bound on the surface motion.

In considering the more spectacular permanent ground displacements that can
result from ground shaking, it should not be forgotten that elastic displacements
also occur and are critical in the design of piles, underground pipelines and

Fig. 1.2 A fault passed through the Shinkhang dam, Taiwan, in 1999, causing the vertical displacement of 9 m

culvert-type structures. Failures in underground piping and ductwork are common and have important implications for post-earthquake emergency services.

1.3 Structural collapse

Figures 1.3–1.11 show some of the many ways in which structural collapse can occur in buildings. Collapse can initiate at any level and may be due to lateral

Fig. 1.3 Total collapse of a multi-storey reinforced concrete structure in Mexico City, 1985

Fig. 1.4 Ground-storey (soft-storey) collapse of buildings in Erzincan, Turkey, 1992

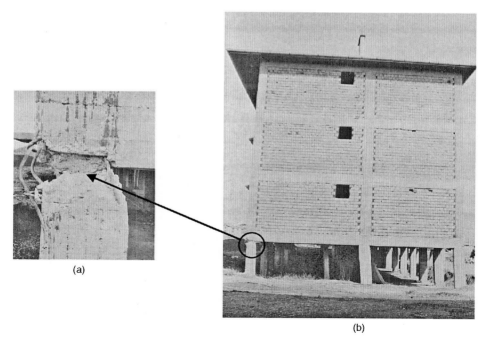

(a)

(b)

Fig. 1.5 Incipient soft-storey collapse of a building in Erzincan, Turkey, 1992

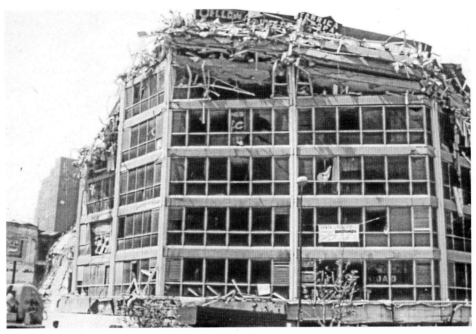

Fig. 1.6 Upper-storey collapse of a multi-storey reinforced concrete structure in Mexico City, 1985

Fig. 1.7 Intermediate-storey failure of a multi-storey structure in Mexico, 1985, probably caused, or aggravated, by buffeting against the adjacent building

Fig. 1.8 The nearer end of this building is restrained by stiff shear walls but the far end was supported by slender columns. The resulting torsional movement has led to collapse in Mexico City, 1985

Fig. 1.9 Total collapse of weak random rubble masonry housing in Gujarat, India, 2001

Fig. 1.10 Failure of modern clay brick house in Erzincan, Turkey, 1992

Fig. 1.11 Failure of multi-storey steel-framed buildings, Mexico City, 1985

or torsional displacement, local failure of supporting members, excessive foundation movement and occasionally the impact of another structure.

An important category of building failure is the case where the building is so badly damaged that, although it has not collapsed, it has to be demolished. For the owner and the insurance company the costs are similar whether the building collapses or is demolished. For the occupants it is the difference between life and death.

1.4   Important categories of damage

Where two buildings are close, or where there is a movement joint, the two sides are likely to pound against each other during an earthquake. Major structural damage can result from this (Fig. 1.7), particularly where the floor levels differ. The cause lies in the closeness of the two structures and in the flexibility of the buildings, both of which are under the control of the designer.

Appendages to buildings – masonry parapets, penthouses, roof tanks, cladding and cantilevers – tend to behave badly (Fig. 1.12). The reasons for this are twofold:

Fig. 1.12   Parapet failure, Gujarat, India, 2001

Fig. 1.13   Earthquake damage to internal fixings has caused batteries to fail in a hospital in San Fernando, USA, 1971

first, many of them are designed without any ductility; and second, the effects of dynamic amplification by the building to which they are attached may greatly increase the forces applied to them.

The contents of buildings often suffer major damage even when the building itself is relatively unharmed. This effect is greater for more flexible buildings and represents an additional reason for the designer to exercise close control over displacements. In many modern buildings the contents are of greater value and importance than the building itself. The costs of preventing damage are often trivial – steel angle ties to the tops of racks, floor bolts to shelving for example. The consequences of failure can be devastating; the batteries in Fig. 1.13 were unable to provide an emergency power supply in a hospital when the mains supply had failed in an earthquake.

Modern buildings are often assembled from many separate components. Older buildings also commonly have timber floors with joists that are poorly tied to the supporting walls. Any lack of tying together in a building is quickly exposed by earthquake shaking (Fig. 1.14). The random nature of earthquake ground motion inevitably leads to differential movement between separate components, and in the absence of structural continuity differential movement will occur.

Anchorages of components into masonry or concrete by cast-in or expanding head bolts are almost invariably brittle in shear and tension, and thus unable to accommodate any movement. Accordingly failures are commonplace, aggravated when the masonry or concrete into which the anchorage is placed is damaged.

Fig. 1.14 Failure to tie in floor joists to walls probably precipitated failure in this housing in the Friuli, Italy earthquake, 1980

1.5 Reinforced concrete

Buildings consisting of frames built from reinforced concrete beams and columns and which are not braced by walls have proved very vulnerable to earthquakes, unless special design and detailing measures are in place to resist earthquakes. The main points of vulnerability are the following

(*a*) beam–column joints (Fig. 1.15)
(*b*) bursting failures in columns (Fig. 1.16)
(*c*) shear failures in columns (Fig. 1.17)

Fig. 1.15 Failure of a beam–column joint in Erzincan, Turkey, 1992. Note failure of the concrete in the joint, and bursting out of column steel

Fig. 1.16   Bursting failure in a column, Northridge, California, 1994. Inadequate horizontal steel has caused the heavy main bars to buckle, and allowed the concrete in the column to shatter

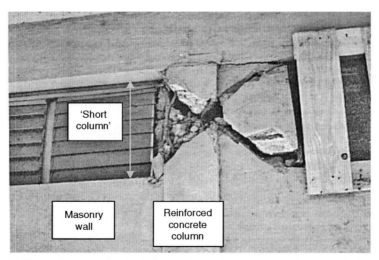

Fig. 1.17   Shear failure of a lightly reinforced concrete column and the adjacent masonry, opposite a window opening in St Johns, Antigua, 1974. The masonry wall stopping short of the top of the column creates a 'short column' liable to fail in shear before bending

Fig. 1.18 Survival of precast panel buildings adjacent to razed precast frame buildings, Spitak, Armenia, 1988

(*d*) anchorage failure of main reinforcing bars in beams and columns (Figs 1.15 and 1.20).

Some or all of these defects may contribute to the ubiquitous 'soft storey' failures referred to earlier (Figs 1.4 and 1.5).

Buildings with shear walls providing a significant contribution to lateral resistance have proved much less vulnerable. Perhaps the most dramatic example was in the Spitak Armenia earthquake of 1988, where all the precast concrete frame buildings suffered total collapse, while the precast concrete wall buildings, although equally poorly constructed, survived without endangering their occupants (Fig. 1.18). Shear walls however do suffer, particularly in compression failure of their outer edges (Fig. 1.19) and in diagonal shear at their bases (Fig. 1.20).

1.6 Structural steelwork

Structural steel shows the following types of damage from earthquakes

(*a*) brittle failure of bolts in shear or tension
(*b*) brittle failure emanating from welds
(*c*) member buckling, including torsional buckling
(*d*) local web and flange buckling
(*e*) uplift of braced frames

Fig. 1.19   Compression failure of outside edge of shear wall, Erzincan, Turkey, 1992

Fig. 1.20   Collapse of tall shear wall building, Baguio, Philippines, 1990

(*f*) local failure of connection elements such as cleats and Ts

(*g*) bolt slip

(*h*) high deflections in unbraced frames

(*i*) failure of connections between steel members and other building elements, such as floors.

1.7 Masonry

Failure of unreinforced masonry is so common that it is almost taken for granted and forgotten (Figs 1.9–1.10). US earthquake codes ban the use of unreinforced masonry altogether in earthquake country, although European codes allow low-rise unreinforced masonry housing provided certain stringent conditions are met. However, economic reasons still ensure that it is widely used both for low-rise structural walls and as infill to framed structures.

Different forms of masonry construction perform in different ways. Weak masonry walls formed from adobe (baked mud) or from random rubble (stones randomly set in weak mortar) perform the worst, and 100% destruction of such buildings is quite common in a severe earthquake (Fig. 1.9); this type of construction is commonly built by the occupiers in poorer areas. Buildings formed from regular, dressed blocks of good-quality granite or other rock are much less likely to suffer total collapse, although there may be damage (Fig. 1.12); this form of construction is found in large official buildings. The ability of good-quality masonry to remain (just) stable after experiencing major cracking and deformation is frequently amazing (Fig. 1.21).

Fig. 1.21 Damage to a palace after the Gujarat, India earthquake of 2001

Fig. 1.22 Out-of-plane failure of an unreinforced, unsecured masonry panel, Antigua, 1974

Failures of both reinforced and unreinforced masonry in-plane are common. In-plane, masonry is very stiff, so that the forces transmitted by ground shaking are high; masonry is also brittle so that failure is accompanied by a marked reduction in strength and stiffness. Damage normally comprises either collapse or diagonal cracking in both directions ('X' cracking) and damage will often be worse around openings.

Horizontal reinforcement laid in the mortar bed joints substantially increases in plane strength and ductility, as does vertical reinforcement in mortar columns cast into hollow blocks. Introducing masonry tie-beams and columns into masonry walls to form smaller confined panels also substantially improves in-plane resistance.

Out-of-plane, free-standing masonry or masonry that has separated from any adjacent structure is liable to toppling failure (Fig. 1.22). Toppling is much less likely if some mechanical connection exists at the sides and head of the wall. Reinforcement continued into a surrounding frame is most effective in avoiding complete collapse, acting as a basket to the masonry even when it is severely damaged.

1.8   Timber

Timber has a generally good record of earthquake resistance because it possesses good tensile strength (though generally limited ductility) and a favourable strength-to-weight ratio. However, it is prone to fungal decay and attack by parasites, which can effectively reduce seismic resistance to zero. Major fires have broken out after earthquakes in cities with timber houses, most notably in Tokyo after the 1923 earthquake, but also more recently for example in the Marina district of San Francisco in 1989 and Kobe, Japan in 1995.

1.9 Foundations

Failure of spread foundations is usually the result of failure of the supporting soil, which is often associated with liquefaction (in which loose, saturated, granular soils effectively turn to quicksand under earthquake shaking, and lose their shear strength). Often, these failures result in gross settlements, but the failing soil is unable to transmit strong shaking to the structures which survive (Fig. 1.23).

Piles are susceptible to failure at their junctions with the superstructure (Fig. 1.24) and where they pass through the junction between soils of differing stiffness

Fig. 1.23 Gross settlements due to liquefaction, Dagupan, Philippines, 1990

Fig. 1.24 Failure at pile-superstructure in a bridge, Loma Prieta earthquake, California, 1989

(for example from alluvial material into rock), because of the large horizontal shearing deformations the soils experience at these points. Junctions between liquefying and non-liquefying soils are a particular case in point.

1.10 Non-structural elements

At any level on a multi-storey building the ground motion will be modified by the motion of the building itself. Generally the effect is to concentrate the frequency of response around a band close to the natural frequency of the building and to amplify the peak acceleration roughly in proportion to the height, reaching an amplification of perhaps two or three at roof level. For any contents which are either very stiff or which have a natural frequency of their own close to that of the building, this means that they are subjected to greater forces than they would be if mounted at ground level.

Experience shows that non-structural items which are suspended, such as ceiling systems and light fittings, perform badly. Appendages such as parapets also suffer high levels of damage (Fig. 1.12). Damage also increases on multi-storey structures towards the roof so that roof tanks and penthouses are subjected to particularly high forces.

Mechanical and electrical systems generally survive earthquake shaking quite well, provided they are well anchored to the main structure. It is generally when these anchorages fail that problems occur (Fig. 1.13).

These failures are due to inertia forces caused by high accelerations. Failures can also occur due to relative displacements. For example, windows and cladding elements are frequently connected rigidly to more than one level and, if there is no ductile provision for movement in the connections, they will fail. Services crossing structural joints, or where they emerge from the ground to enter a building, are similarly subjected to relative displacements which may damage them. Walkways between buildings may be particularly vulnerable.

1.11   Bibliography

Earthquake reconnaissance reports are prepared and published by a number of sources. In the UK, EEFIT (Earthquake Engineering Field Investigation Team) has published regular earthquake reports since 1983, and these are available together with photos of earthquake damage from its sponsoring organisation, the Institution of Structural Engineers (www.istructe.org.uk). Imperial College, London (www.ic.ac.uk) also has a long history of earthquake reconnaissance missions for which reports are available. Elsewhere, the Earthquake Engineering Research Institute, based in California (www.eeri.org) similarly has been publishing excellent earthquake reconnaissance reports for many years, and posts summaries on its website within weeks of the events occurring. The New Zealand Society for Earthquake Engineering (www.nzsee.org.nz) is another good source of English-language reports.

2 Ground motion

'What to do in an earthquake? Stand still and count to one
hundred. By then it will be over.'

Professor George W. Housner

This chapter covers the following topics.

- The nature of ground motion
- The principal factors in assessing ground motion
- Influences on ground motion
- Means of describing ground motion
- The design earthquake

2.1 Primary and secondary sources of earthquake damage

The potential for a large earthquake to cause damage comes in the first instance
from the violent shaking of the ground, which may affect an area many hundreds
of kilometres in radius. This is the primary source of damage with which
earthquake engineers must deal. Large relative displacements across a fault
which breaks up to the surface can also be damaging but usually relatively few
unlucky structures are affected. For example, about 100 000 buildings collapsed
or were severely damaged in the Kocaeli, Turkey earthquake of 1999, which
occurred on a fault over 100 km long with displacements averaging 3–4 m, but
of these buildings only about 100 directly straddled the fault break.

The seismic shaking causes direct effects on structures, due to the inertia forces
set up by the ground accelerations, but important secondary sources of damage
may also arise. Most significantly, large soil movements may occur due to consolidation, liquefaction (the temporary loss of shear strength in loose, saturated,
sandy soils), landslides or avalanches. Coastal sites may need to consider tsunamis
(commonly referred to as tidal waves) triggered by offshore earthquakes; they are
discussed further by Synolakis (2003). Other secondary effects are those due to fire
following an earthquake (which has cost many lives in the past, see Scawthorn,
2003), the collapse of one structure onto another, and the release of noxious
chemical or radioactive materials, although to date the last possibility has never
caused major problems.

The remainder of this chapter concentrates on the primary hazard of
strong ground motion, and how to describe it for the purposes of engineering
design.

2.2 Earthquake basics

2.2.1 Earthquake sources

Figure 2.1 shows some of the principal terms used in describing an earthquake's location.

Earthquakes arise due to forces within the earth's crust tending to displace one mass of rock relative to another. When these forces reach a critical level, failure in the rock occurs at points of weakness called fault planes and a sudden movement occurs, which gives rise to violent motions at the earth's surface. The failure starts from a point on the fault plane called the focus, and propagates outwards until the forces in the rock mass are dissipated to a level below the failure strength of the rock. The fault plane may be hundreds of kilometres long in large earthquakes, and tens of kilometres deep. In a large earthquake, the fault plane is likely to break up to the surface, but in smaller events it remains completely buried. A more complete description of the causes and types of earthquake is given by Bolt (2001).

2.2.2 Quantifying earthquakes

There are two principal measures of an earthquake. Earthquake *magnitude* is a fundamental property of the earthquake, related to its energy release on a logarithmic scale. By contrast, earthquake *intensity* describes the effects of the earthquake on the Earth's surface, by observing its effects on people and buildings. Unlike magnitude, the intensity of a given earthquake depends on the location at which it is measured; in general, the larger the epicentral distance (see Fig. 2.1) the lower the intensity. Thus a given magnitude of earthquake will give rise to many different intensities in the region it affects. It is important to recognise this fundamental distinction between the two measures.

A number of different magnitude scales exist. Two common scales are the body wave magnitude m_b (suitable for measuring smaller magnitude events) and the surface wave magnitude M_s (most suitable for large events). Both are measured from sensitive instruments (seismographs) which detect ground tremors at great distances from the earthquake source. A third scale is M_w, the moment magnitude.

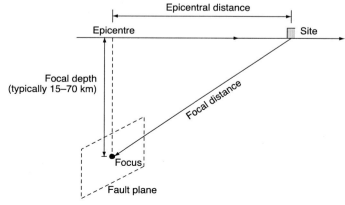

Fig. 2.1 Definitions of earthquake sources location

This is directly related to the estimated energy release at the earthquake source and is suitable for all sizes of event. In broad terms, an earthquake with magnitude less than 4 on any of the scales is unlikely to cause significant damage, while magnitudes larger than 8 are rare events affecting very large areas. Because of the logarithmic nature of the scale, a one point increase in magnitude represents a thirtyfold increase in energy release, and earthquakes larger than 9.5 are not found in practice because they would represent fault sizes larger than the dimensions of the Earth's crust.

Intensity scales rely on reports of the felt and observed effects of an earthquake at a given position. Since this is less precise than the measurement of magnitude, intensity values are described by Roman numerals. A number of different scales exist. Among these are the 12-point Modified Mercalli Intensity (MMI) scale which is commonly used in the USA; intensity I is the lowest (not felt except by a few under especially favoured conditions), VII is the intensity at which some structural damage is likely and XII the highest (total damage). The European Macroseismic Intensity (EMS) scale (a development of the MSK scale) is another roughly similar 12-point scale. It is more favoured in Europe, since it relates damage more precisely to the earthquake-resisting qualities of the damaged structures. The Japanese Seismic Intensity scale is similar in principle but is based on only seven points.

Clearly, intensity and magnitude are related to some extent, in that in general larger magnitudes give rise to larger intensities for a given epicentral distance. Ambraseys (1985) provides relevant data for north-west European earthquakes.

2.2.3 Occurrence of earthquakes

Table 2.1 shows the number of earthquakes that occur on average per year, as a function of magnitude. Of course, many of the earthquakes are remote from human populations and cause little, if any, damage. Figure 2.2 shows the number of fatalities caused by earthquakes for each decade of the twentieth century; it can be seen that there has been little if any progress in reducing casualties. However, the world population tripled during the century, and so the number of people at risk from earthquakes also increased, especially in megacities which have proved particularly vulnerable.

Figure 2.3 places earthquake risk in the context of that from other natural hazards. It can be seen that wind storms and floods claimed more lives than earthquakes during the second half of the twentieth century. However, earthquakes

Table 2.1 Annual numbers of earthquakes worldwide

Magnitude M_s	Average number per year $>M_s$
8	1
7	20
6	200
5	3000
4	15 000
3	>100 000

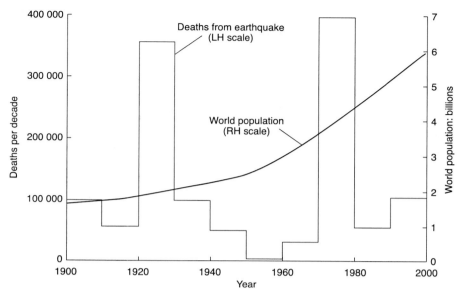

Fig. 2.2 Estimated fatalities from earthquakes during the twentieth century

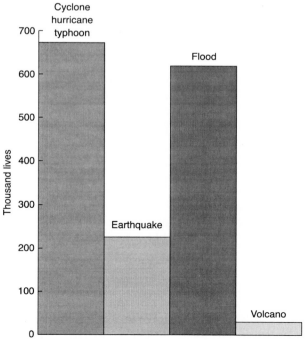

Fig. 2.3 Lives lost by type of natural disaster, 1947 to 1991 (Wassef 1993)

have the potential to cause the largest impact in terms of lives lost and economic loss due to a single event. It is this risk from rare but extreme events which make earthquakes particularly challenging to design for, an issue which is returned to in subsequent sections.

Details of earthquakes are posted on a number of websites more or less as they occur worldwide; the British Geological Survey (www.earthquakes.bgs.ac.uk) and the US National Earthquake Information Center (NEIC) (earthquake.usgs.gov/recenteqsww/quakes/quakes-all.html) are examples. Lists of past earthquakes (including those before instrumental records were available) are also freely available, for example from the NEIC on neic.usgs.gov/neis/epic. A useful discussion of earthquake catalogues is provided by Musson (2005).

2.3 Earthquake probability and return periods

Almost anywhere in the world is thought to be susceptible to suffering an earthquake of magnitude 6, which can give rise to very severe motions at its epicentre. However, in an area of low seismicity such as the UK, such an occurrence would be extremely rare, and it would be unreasonable to design against it, except perhaps for high-risk installations such as nuclear power stations. Therefore, at any rate in areas of low seismicity, something less than the 'maximum credible event' must be found for design, because the magnitude 6 event at a given site is very likely to be credible, although it may be extremely rare. The 'maximum credible' concept is more useful for sites near large active faults which break regularly within human time spans of tens or hundreds of years; here, the worst that could occur may be much better defined, and may need checking. However, there is now a general consensus that a *probabilistic* approach to defining earthquake hazard gives the most appropriate results for engineering design. Thus, the design earthquake is defined by its annual probability of exceedence P_1, or (equivalently) its return period T.

P_1 is defined as the probability in any given year that ground motions of a given intensity will be exceeded. For example, in parts of California there is a 2% annual probability that ground accelerations exceeding 0.25 g may occur, while in the UK the probability is likely to be nearer 0.01%. The return period T is then defined simply as

$$T = 1/P_1 \tag{2.1}$$

Often, the probability of exceedence P_y during a period of y years (i.e. greater than annual) is of most interest, where y years might represent the lifetime of a building. P_y, T and P_1 are related as follows

$$P_y = 1 - (1 - P_1)^y = 1 - \left(1 - \frac{1}{T}\right)^y \tag{2.2}$$

For example, a 475-year return period corresponds to a probability of exceedence in a 50-year building life of $1 - (1 - 1/475)^{50} = 10\%$. Generally, design return periods of the order of 500 or more years are used in earthquake-resistant design, rather than 50 years as used for many other environmental loads such as wind, because the 'tail' of the earthquake hazard distribution – the effect of rare but extremely damaging events – is generally much more significant for earth-quakes than for wind. Therefore, relatively much rarer events must be considered, in order to reduce the risk of failure to levels comparable to those of other hazards. In low-seismicity areas, this is particularly true, because of the almost limitless upper bound to the 'maximum credible event' discussed above.

2.4 Performance objectives under earthquake loading

At any rate in principle, more than one level of design needs to be considered. Current US definitions of performance goals developed in the USA for rehabili-tating existing buildings are defined as follows (FEMA 2000). Similar definitions are given for the design of new buildings (FEMA 1997).

(a) Operational. Minimal or no damage to structure and non-structure; backup utility services maintain functions.
(b) Immediate Occupancy. The building remains safe to occupy; any repairs are minor, although some non-structural systems may not function, either because of lack of electrical power, or internal damage to equipment.
(c) Life Safety. Structure remains stable and has significant reserve capacity; hazardous non-structural damage is controlled. However, repairs may be required before the building can be reoccupied, and such repairs may not prove economic. The risk to life safety is low.
(d) Collapse Prevention. The building remains standing, but only barely; any other damage or loss in acceptable. There may be a significant threat to indi-vidual life safety, but since the building avoids collapse, gross loss of life should be avoided. Many buildings meeting this goal will be complete economic losses.

Clearly, as the performance goal become less stringent (i.e. changing from Operational to Collapse Prevention), the return period of the earthquake for which the goal has to be met can become longer. FEMA (1997) sets indicative earthquake return periods for new buildings as shown in Table 2.2.

These goals may be appropriate for most buildings, but some will require enhanced performance goals. For example, hospitals (particularly those involved in treating acute cases during and after an earthquake) need a greater level of functionality, and the objectives of Table 2.3 might typically apply.

Conversely, an existing warehouse building containing non-hazardous material which is due for demolition in a few years time might be upgraded merely to meet a 'Collapse Prevention' performance goal with a 2% chance of being exceeded during its remaining life.

Table 2.2 Performance objectives for normal buildings

Performance goal	Return period
Immediate Occupancy	~75 years (50% in 50 years)
Life Safety	~475 years (10% in 50 years)
Collapse Prevention	~2475 years (2% in 50 years)

Table 2.3 Enhanced performance objectives for hospitals

Performance goal	Return period
Operational	∼75 years (50% in 50 years)
Immediate Occupancy	∼475 years (10% in 50 years)
Life Safety	∼2475 years (2% in 50 years)

2.5   Representation of ground motion

2.5.1   Earthquake time histories

The earthquake intensity described above gives a broad measure of the damaging power of an earthquake at a given location, but more precise (and less subjective) measures are required by engineers for the purposes of design calculations. The most precise description is given by a 'time history' of the motions at a given point. Time histories are measured by strong motion accelerographs set into action by the earthquake itself when the ground acceleration exceeds a preset threshold. Digitised records of earthquakes are available from a number of sources; the PEER strong motion database in California (http://peer.berkeley. edu/smcat/) and the European strong-motion database (http://www.isesd.cv.ic. ac.uk/) are examples. Further databases are discussed by Bommer and Strasser (2004).

Because the duration of an earthquake is short – 60 s is a long record – strong motion records contain little information about the very low frequency components of ground motion. It should also be borne in mind that strong-motion records are commonly taken in locations in or near heavy buildings or engineering structures which have some filtering effect, biasing the frequency content.

Figure 2.4 shows plots of horizontal acceleration, velocity and displacement against time for a Californian earthquake. The acceleration plot is a record from a strong-motion accelerograph and the other two plots have been obtained from it by integration. This is the simplest type of record and provides precise information about one specific earthquake. Important parameters associated with time-based records are the peak values and the duration of strong motion. For computation, earthquake records with digitised values at intervals of around 0.005–0.01 s are commonly used.

2.5.2   Earthquake response spectra

The time-history plots, though they contain a great deal of information, suffer from two disadvantages. First, it is difficult to judge what the frequency content is, and hence the damaging power for structures of varying natural periods of vibration. Second, they are specific for a given time and place, and would not be repeated even at the spot in a subsequent earthquake. The earthquake response spectrum provides major advantages in both respects, and represents a most useful tool for a design engineer to characterise earthquake motions. It represents the peak response of a linear elastic, single degree of freedom spring-mass-damper structure to an earthquake, plotted against the structure's natural period. Contin-

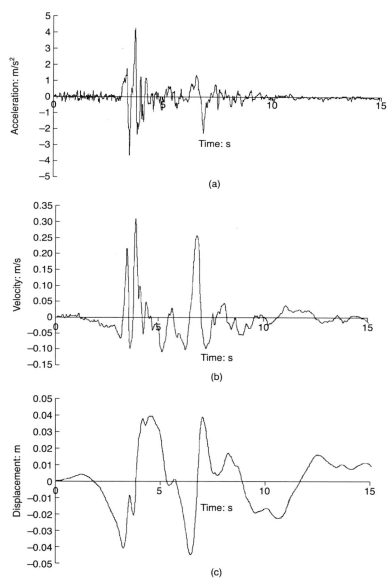

Fig. 2.4 Time-history plots for a record from the Northridge, California earthquake of 1994: (a) acceleration; (b) velocity; and (c) displacement

uous plots are drawn for each value of damping selected so that a response spectrum is represented by a family of curves, as illustrated in Fig. 2.5.

Response spectra are discussed in more depth in Chapter 3. However, it is immediately clear from the peaks in the spectra shown in Fig. 2.5 that the Northridge ground motions would have been particularly damaging to structures with a natural period in the range 0.1–0.4 s (representing low-rise construction up to

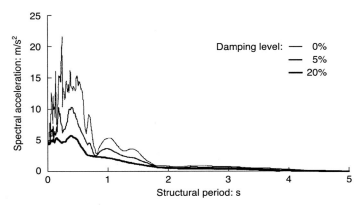

Fig. 2.5 Response spectra for the Northridge, California earthquake ground motions of Fig. 2.4

around five storeys) with damping levels between 2% and 5% (a typical range for buildings responding at around their yield capacity). Increasing the damping level to 20% would have been highly effective in reducing the response in low-rise construction; this level of damping is a possibility with the addition of special damping elements, as discussed in Chapters 13 and 14. These observations would have been difficult to make from the time-history traces of Fig. 2.4.

A further advantage of a response spectrum is that by averaging and enveloping the spectra of several related time histories, thus smoothing the individual peaks and troughs, a representation of ground motion may be obtained which is more general than can be obtained from a particular time history (Fig. 2.6). Such smoothed spectra are almost always the basis for spectra used in design.

Representing ground motions by their response spectra therefore overcomes two of the disadvantages of time-history representations. They give some immediate information about structural response, and they can be generalised to cover the effects of a range of possible earthquakes at a given site.

Other forms of spectral representation of ground motion are possible, in particular Fourier spectra and power spectral density plots. These have more specialist uses for the earthquake engineer, and the reader is referred to Mohraz and Sadek (2001).

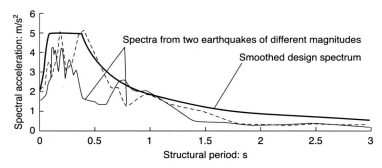

Fig. 2.6 Smoothed design spectrum

2.5.3 Damaging capacity of ground motions

Peak ground acceleration (pga) is often quoted as the single figure representing the severity of a particular earthquake motion. Usually it is quoted as a percentage of the acceleration due to gravity; multiplying by 9.8 converts to m/s^2. In itself, pga is rather unreliable as an indicator of damaging capacity because it may occur as only the briefest of transient values and gives no information on the frequency content of the motions. Peak velocity gives some information about longer period content within the ground motion, and is likely to be better correlated to damage (Walde *et al.*, 1999). A response spectrum, as discussed above, gives even more information about the effect of the motions on structures with varying structural periods and damping levels. Even here, the information relates to linear elastic structures which presupposes little or no damage. 'Ductility modified' response spectra relate to yielding structures, and are discussed in more detail in Chapter 3.

2.5.4 Vertical, torsional and rocking components

For most structures the horizontal translation of earthquake ground motion has by far the greatest effect. However, four other components – one vertical translation and three rotations – also exist and may need to be explicitly taken into account in some cases.

Vertical motions affect long-span structures, since they may significantly increase bending and shear forces due to gravity loads. They may also reduce the effect of gravity loads in maintaining overall stability against lateral loads, and so should also be allowed for in stability calculations. In the past, the customary assumption was that the peak vertical acceleration was two-thirds of the peak horizontal acceleration and had a similar spectral distribution. However, this is a crude approximation; on firm ground near the epicentre of earthquakes, the vertical motions can be much greater than the horizontal ones in the short period range, while they become relatively insignificant far from the epicentre. Codes such as Eurocode 8 (CEN 2004) and ASCE 4 (1998) allow for this. Moreover, on soft soil sites, the vertical motions are likely to be amplified much less than the horizontal ones, because the vertical compressive stiffness of the soil is usually greater than its shear stiffness, so that vertically propagating waves pass through more or less unmodified. As explained in the next section, analytical methods exist for estimating the way soils modify ground motions, so if vertical motions are important, the soil effects should be accounted for separately for vertical motions.

Rotational ground motions are less likely to be important and are seldom allowed for. Rocking motions (i.e. rotation about a horizontal axis) may affect very tall slender structures, and Eurocode 8 Part 6 requires this to be accounted for in tall masts and chimneys, supplying a suitable rocking spectrum. Torsional ground motions (i.e. rotation about a vertical axis) in themselves are unlikely to need inclusion. However, the coupled torsional–horizontal response of torsionally unbalanced structures can be very troublesome. These are triggered by horizontal (translational) shaking, so no special allowance needs to be made in specifying the input motions.

2.6   Site effects

2.6.1   Soil amplification of ground motions

It has long been observed that buildings (particularly tall ones) founded on poor soil generally perform much worse in earthquakes than those founded on hard soil or rock. A notable example occurred during the Mexican earthquake of 1985. This magnitude 8.1 event had its epicentre just off the Pacific coast, and by the time the earthquake motions had travelled approximately 300 km eastwards to Mexico City, they had attenuated to such an extent that they were not particularly damaging. However, the centre of Mexico City is built on 30 m of very soft clay which acted as a resonator for the motions, and amplified them in the same way as would a wobbling jelly. A high proportion of buildings, particularly of 10–20 storeys, were severely damaged in this soft-soil region, while very few were damaged on firm ground.

Figure 2.7 compares the response spectrum of motions on hard ground at Mexico City with those on the soft soil. The figure clearly shows the transformation of essentially harmless motions into ones that were particularly damaging to structures within a period of approximately 2 s.

Mexico City is an extreme example; the surface soils there are unusually soft, possess little damping but relatively high strength and there is a large contrast in stiffness with the material below the soft clays. However, some degree of amplification can almost always be expected in the presence of soft soils, and it must be accounted for in design. Figure 2.8 shows the design response spectra for different soil types given in Eurocode 8 (CEN 2004). It can be seen that there can be a particularly significant effect for structural periods around 0.7–1 s. Since people tend to settle on soft alluvial basins, because such basins are fertile and can support large populations, the problem is clearly a widespread one.

The soil amplification effects are most marked for low-amplitude shaking. However, as the shaking intensity increases, many soft soils begin to yield, which tends to trim the motions that reach the surface. The seismic sections of

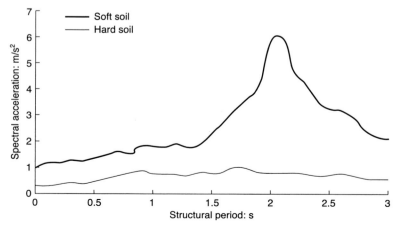

Fig. 2.7 Graph showing 5% damped spectra for Mexico City, 1985

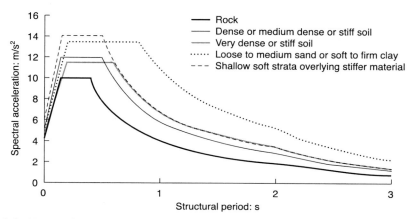

Fig. 2.8 Eurocode 8 elastic ground response spectra for 5% damping

the US code IBC: 2003 (ICC 2003) allow for this effect, but the 2004 edition of Eurocode 8 does not.

2.6.2 Soil–structure interaction
The presence of soft soils at a site may have other effects on structural response, beyond amplifying motions. First, there may be significant flexibility introduced by the restraint the soft soils offer the structures they support, which will tend to increase the natural period of the structures. Generally, Figs 2.5 and 2.8 suggest that an increase in period will reduce acceleration response, except for very stiff, short-period structures. However, for the unusual spectrum of Fig. 2.7, lengthening the period of structure, which would have been (say) 1.5 s on a rigid soil, will in fact increase response. Also, although accelerations and hence forces typically reduce due to foundation flexibility, deflections are likely to increase, which may increase the risk of impact between adjacent structures and will also increase the P–delta effects discussed in Chapter 3.

Another possible effect applies to massive structures like dams or nuclear containment structures which may be sufficiently large and stiff to modify locally the ground motions in soft soils. Both these effects – modification of foundation flexibility and local ground motions – are known as soil–structure interaction (SSI) effects and are further discussed in Chapter 3.

2.6.3 Topographical effects
Another kind of amplification is found to occur at or near ridges or sharp changes of slope. Eurocode 8 Part 5 Annex A recommends that such effects should be accounted for in sites near long ridges and cliffs with a height greater than about 30 m and provides for increases of up to 40%. Amplifications are also found near the edges of some deep alluvial basins, although these have yet to be quantified in codes of practice; Faccioli (2002) provides further information.

2.7 Quantifying the risk from earthquakes

2.7.1 World seismicity

Earthquake activity is not uniformly spread across the Earth; in fact, 90% of energy release from earthquakes occurs on the 'ring of fire' around the edge of the Pacific ocean, with another active belt stretching across the southern edge of Europe to the Himalayas and into Eastern China. Figure 2.9 shows earthquakes of magnitude greater than 5 recorded in the period 1980–1989. The earthquakes are mainly concentrated along relatively narrow bands at the junctions between the tectonic plates which together form the strong but brittle outer skin of the Earth, called the lithosphere. Although earthquake occurrence is strongly concentrated at the plate boundaries, there are also many areas of more distributed seismic activity remote from the plate boundaries. Also, as noted on Fig. 2.9, some boundaries, particularly at the southern edge of the Eurasian plate, are rather indistinct.

2.7.2 Probabilistic estimates of ground motion

As discussed in section 2.3, earthquake engineers need information on seismic ground motions at a site with a specified return period, which often needs to be around 500 years or longer. A statistical analysis can be made of the location and magnitude of earthquakes in the region surrounding a site, which can then be used to quantify the seismic hazard at a site, provided it is combined with a knowledge of how ground motions attenuate with distance. For example, the response spectrum with a 475-year return period (10% probability of exceedence in 50 years) can be calculated. A response spectrum plotted so that the exceedence probability is equal for all structural periods is called a uniform hazard spectrum. An alternative representation of hazard takes the form of a plot of a ground motion parameter, such as peak ground acceleration or velocity, against return

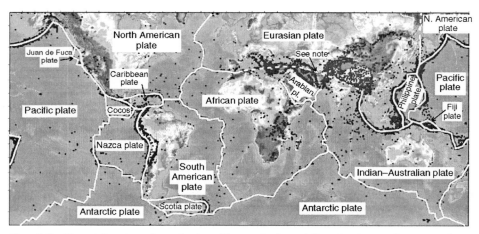

Fig. 2.9 Tectonic plate boundaries. Note: A dotted line is drawn from southern Europe through to China because the earthquake belt in that region occurs over a wide distributed zone of deformation, and a distinct plate boundary does not apply

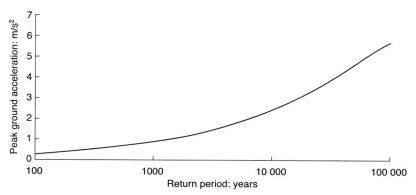

Fig. 2.10   Seismic hazard curve on rock sites for South Wales, UK (Arup 1993)

period (Fig. 2.10). The methods of carrying out such probabilistic estimates are described in a definitive monograph by Maguire (2004).

Probabilistic hazard assessments depend on the availability of reliable catalogues of earthquakes. Since the return periods of interest are long, instrumental records of earthquakes, which are not available before about 1918, may be insufficient, particularly in areas of low seismicity where the hazard may be dominated by very rare events. To fill this gap, reports of earthquakes in newspapers, monastic records and other historical sources can be useful; the severity and extent of the damage can be used to estimate the location and magnitude of the earthquakes which caused it. Such historical reports go back to periods as early as 1200 BC in some parts of the world and form an important part of our knowledge of seismicity. Geological studies also provide important clues to past activity, especially where earthquake sources are shallow and faults can be identified on the Earth's surface.

In assessing the seismicity of a site all the available information needs to be considered. Once this has been done, it should be remembered that any seismicity prediction remains an estimate with a substantial degree of uncertainty.

2.7.3   Published sources of hazard
Producing a seismic hazard curve such as that shown in Fig. 2.10 is a lengthy exercise, which requires specialist expertise beyond the scope of most earthquake engineers. In most cases, however, this exercise is not required. Reliable published maps of seismic hazard distribution are now readily available. In particular, the Global Seismic Hazard Assessment Program (GSHAP) has published authoritative maps of 475-year peak ground accelerations for the whole world on http://seismo.ethz.ch/GSHAP/; Fig. 2.11 gives an indication of the spread of worldwide seismicity that it reveals. The GSHAP project was completed in 1999, and so contains relatively up-to-date information.

Seismic codes of practice also generally contain seismic hazard maps, which can be found for many countries in the World List published by the International Association for Earthquake Engineering (IAEE 2000) which is updated every four years. Although the level of hazard indicated in these maps is likely to be a

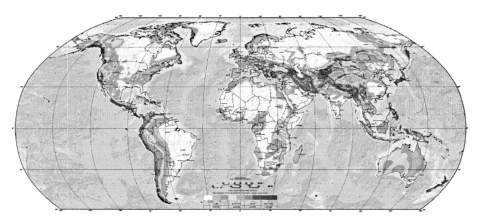

Fig. 2.11 GSHAP global seismic hazard map (GSHAP 1999)

statutory minimum obligation for design purposes, the date of the code needs to be carefully checked because sometimes published seismic hazard maps in codes do not keep pace with the most recent information.

2.7.4 Areas of low seismicity
The public is, naturally enough, greatly concerned with areas of high seismicity. However, in many areas of low seismicity the problem can be dismissed too readily. Few areas of the world are free from earthquakes altogether; even a low-seismicity area such as the UK has a significant risk at long return periods, as Fig. 2.10 shows.

A normal well-constructed building, designed for moderate wind forces, should be able to resist minor ground shaking reasonably well. However, structures which are unusually flexible or exceptionally brittle may be sensitive to small nearby earthquakes or more distant larger earthquakes. The relative importance of earthquake resistance becomes greater for structures of great importance or where the consequences of failure are especially serious, such as nuclear reactors or stores for dangerous chemicals, since longer return periods must be considered in design.

The design earthquake for a sensitive structure in an area of low seismicity is most likely to be a comparatively small-magnitude event occurring close to the site. The ground motion resulting from this characteristically has a high peak acceleration but a short duration. Booth and Pappin (1995) discuss these issues further.

2.8 Design earthquake motions
2.8.1 Response spectra in seismic codes of practice
Response spectra form the basis for defining the earthquake motions in current codes of practice. In most cases, these will form the main basis for design, although they may be supplemented by the special studies described in the next subsection.

Code spectra always take the form of smoothed spectra intended to envelope the range of conditions for which they are specified (Fig. 2.6), and are supplied for a range of standard soil profiles (Fig. 2.8). The spectra are usually normalised by the pga (peak ground acceleration) to give a standard shape; that is, the spectral accelerations are divided by the pga. The design spectrum is then found by multiplying the standard normalised shape by an appropriate pga.

Generally, the pga relates to a value on rock, thus removing any amplifying influence of overlying soils. Rock pgas are usually supplied in seismic zoning maps attached to the codes and give a measure of the underlying seismicity of the region. The global seismic hazard maps produced by GSHAP (Fig. 2.11) are of this form. The standard procedure to define a code response spectrum is therefore to chose an appropriate pga for the site and a standard soil profile most appropriate for the site. The spectral acceleration is then given as a function of structural period. Eurocode 8 uses this system; an additional choice has to be made of whether the hazard at the site is dominated by large magnitude (Type 1) earthquakes (often the case for areas of high seismicity) or smaller magnitude (Type 2) earthquakes (more typical for low-seismicity areas). Also in Eurocode 8, the presence of topographical effects (subsection 2.6.3) can optionally be allowed for.

By contrast, US seismic codes after 1997 changed from relating design spectra to pga. Instead, two parameters are used, namely peak spectral acceleration at short period (S_s) and spectral acceleration at 1-second period (S_1). This two-parameter system allows spectral shapes to reflect varying regional geology much better than the use of a single parameter, namely pga. Eurocode 8 may well follow the Americans and adopt a multi-parameter system in future revisions of the code.

2.8.2   Response spectra from special studies

In some circumstances special studies may be carried out to define a response spectrum for a particular site. Such studies are often performed for high-risk installations such as nuclear power stations, and are specified in some codes (including Eurocode 8 and the US code IBC) for sites where the soils may be unusually prone to amplifying ground motions, such as those at Mexico City.

A number of possibilities exist. One such is where the fundamental seismicity of a site is uncertain, either because of the lack of a reliable seismic zoning map for the region or because such zoning maps that do exist relate to different return periods from those required. In this case, a site-specific hazard assessment (subsection 2.7.2) would be carried out, in order to obtain a value for the pga (for Eurocode 8) or S_s and S_1 (for IBC – see previous section) on rock at the site. Provided that the soil profile at the site is well known and well characterised in the code, a design spectrum can then be obtained directly from the code which allows both regional seismicity and site effects to be accounted for.

Alternatively, the regional seismicity may be well specified, but the effects of the overlying soil may not be adequately allowed for in codes of practice. In this case, the earthquake motions found in a notional rocky outcrop near the site would be well known, and these could then be modified to allow for the overlying soils, using the analytical methods described in Chapter 4.

Of course, if neither regional seismicity nor the overlying effects of the soils were well established, both procedures would be necessary. That is, the rock motions would first have to be found, and then the effect of overlying soils allowed for.

2.8.3   Earthquake time histories for design purposes

An earthquake response spectrum is likely to form the initial basis of design ground motions for most projects. However, response spectra provide no information about the duration of the motion. This may be important where significant changes in soil or structural properties take place with time under repeated cyclic loading. To account for this, and to allow more generally for non-linear effects, earthquake time histories are required, in order to carry out a 'time-history' analysis. There are a number of ways of deriving design time histories for a particular site.

The most direct way is to select the records of real earthquakes with a magnitude and distance appropriate to the seismic hazard at the site. For sites where there is significant amplification of the ground motions in the top layers of soil, this is most satisfactorily done by choosing appropriate records taken for rock sites, and then modifying them analytically for the soil, as described in Chapter 4. Codes of practice such as Eurocode 8 require that at least three records be used, and that the average of the response spectra of the individual records should envelope the design response spectrum for the site given in the code. It is likely that some or all of the records would need to be factored in order to achieve this at all structural periods of interest. Even with such factoring, it may be difficult to find a set of records which envelopes the code spectrum without being unduly conservative at some structural periods. Choice of real time histories for design purposes is discussed by Bommer and Acevedo (2004).

A second method is to generate a set of 'artificial' time histories for which the average spectrum envelopes a specified design spectrum. A number of software packages exist to perform this task (e.g. SIMQKE 1976) and it is easily possible to produce a set of three or four statistically independent records which meet the requirements of a code such as Eurocode 8 or (more stringently and completely) the ASCE Standard ASCE 4-98 (1998). The problem is that the records produced tend to look quite unlike those of real earthquakes, and in particular tend to have many more damaging cycles. This becomes important where every cycle of loading may cause cumulative damage or movement; examples are the onset of liquefaction in soils, the sliding or overturning of retained soils and low-cycle fatigue effects generally in structures. Generally speaking, the use of artificial records will be conservative, compared to real records, but the conservatism may be excessive and a realistic understanding of response may not be possible.

A third way has been developed, which is particularly useful for low-seismicity countries such as the UK, where there are almost no ground motion records of damaging earthquakes at all. In this method, sensitive instruments are used to record the accelerations at the chosen site of many small earthquakes. Since the number of earthquakes occurring increases exponentially with decreasing magnitude, a sufficient number of records can be acquired over a period of a few years even in the UK. The small earthquakes represent breaks over small areas of a fault, which are accurately located. By combining a large number of these records using a 'Green's function' technique, the effect of the break of a much larger area of fault, and hence much larger earthquake can be simulated. The technique is further discussed by Erdik and Durukal (2003), who also present other methods of deriving time histories directly from geophysical parameters.

2.9 References

Ambraseys N. N. (1985). Intensity attenuation and magnitude–intensity relationship for NW European earthquakes. *Earthquake Engineering and Structural Dynamics*, Vol. 13, pp. 733–778.

Arup (1993). *Earthquake hazard and risk in the UK*. Report for The Department of the Environment, London.

ASCE 4 (1998). *Seismic Analysis of Safety-related Nuclear Structures and Commentary*. American Society of Civil Engineers, Reston, VA.

Bolt B. A. (2001). The nature of earthquake ground motion. In: Naeim F. (ed.) *The Seismic Design Handbook*. Kluwer Academic Publishers, Boston.

Bommer J. & Acevedo A. (2004). The use of real earthquake accelerograms as input to dynamic analysis. *Journal of Earthquake Engineering*, Vol. 8, pp. 43–91. Special Issue No. 1.

Bommer J. & Strasser F. (2004). Earthquake engineering on the internet: strong motion data. *SECED Newsletter*, Vol. 18, No. 2, pp. 43–91. Institution of Civil Engineers, London.

Booth E. & Pappin J. (1995). Seismic design requirements for structures in the United Kingdom. In: Elnashai A. (ed.) *European Seismic Design Practice*. Balkema, Rotterdam.

CEN (2004). EN1998-1: 2004. *Design of structures for earthquake resistance. Part 1: General rules, seismic actions and rules for buildings*. European Committee for Standardisation, Brussels.

Erdik M. & Durukal E. (2003). Simulation modelling of strong ground motions. In: Chen W.-F. & Scawthorn C. (eds) *Earthquake Engineering Handbook*. CRC Press, Boca Raton, FA.

Faccioli E. (2002). 'Complex' site effects in earthquake strong motion, including topography. *Proceedings of the 12th European Conference on Earthquake Engineering*. Elsevier Science, Oxford.

FEMA (1997). *NEHRP recommended provisions for seismic regulations for new buildings and other structures*. Prepared by the Buildings Seismic Safety Council for the Federal Emergency Management Agency, Washington DC.

FEMA (2000). FEMA 356. *Prestandard and commentary for the seismic rehabilitation of buildings*. Prepared by the American Society of Civil Engineers for the Federal Emergency Management Agency, Washington DC.

GSHAP (1999). Global seismic hazard assessment program. http://seismo.ethz.ch/GSHAP.

IAEE (2000). *Earthquake Resistant Regulations – a World List*. International Association for Earthquake Engineering, Tokyo.

ICC (2003). IBC: 2003. *International Building Code*. International Code Council, Falls Church, VA.

Maguire R. (2004). Seismic hazard and risk analysis. Earthquake Engineering Research Institute, Oakland, CA.

Mohraz B. and Sadek F. (2001). Earthquake ground motions and response spectra. In: Naeim F. (ed.) *The Seismic Design Handbook*. Kluwer Academic, Boston.

Musson R. (2005). Earthquake engineering on the internet: earthquake catalogues. *SECED Newsletter*, Vol. 18, No. 3, Institution of Civil Engineers, London.

Scawthorn C. (2003). Fire following earthquakes. In: Chen W.-F. & Scawthorn C. (eds) *Earthquake Engineering Handbook*. CRC Press, Boca Raton, FA.

SIMQKE – I (1976). A program for artificial motion generation. Department of Civil Engineers, Massachusetts Institute of Technology (distributed by NISEE, University of California, Berkeley, http://nisee.berkeley.edu/).

Synolakis C. (2003). Tsunami and seiche. In: Chen W.-F. & Scawthorn C. (eds) *Earthquake Engineering Handbook*. CRC Press, Boca Raton, FA.

Walde D. J., Quitoriano V., Heaton T. H., Kanamori H., Scrivner C. W. & Worden, C. B. (1999). PriNet "Shakemaps": rapid generation of peak ground motion and intensity maps for earthquakes in southern California. 15, 537–555.

Wassef A. M. (1993). *Relative impact on human life of various types of natural disaster: an interpretation of data for the period 1947–91*. In: Merriman P. & Browitt C. (eds). *Natural Disasters – Protecting Vulnerable Communities*. Thomas Telford, London.

3 The calculation of structural response

'The myth, then, was that refinement of the analysis process improved the end result.'

Nigel Priestley. *Myths and Fallacies in Earthquake Engineering, Revisited.* The Mallet Milne Lecture 2003, IUSS Press, Pavia, Italy

This chapter covers the following topics.

- Stiffness, mass, damping and resonance
- Linear response spectrum analysis
- Linear and non-linear time-history analysis
- Equivalent static analysis
- Capacity design
- Displacement and force based analysis
- Non-linear static (pushover) analysis
- Target displacement and capacity spectrum methods
- Modelling of building structures

3.1 Introduction

Earthquake loading poses the structural analyst with one of the most challenging problems in engineering. A violent and essentially unpredictable dynamic ground motion imposes extreme cyclic loads on engineering materials whose response under such conditions is complex and incompletely understood. If this is the case, engineering designers for whom this book is written may wonder whether there is any point in their getting to grips with the complex underlying theory of dynamic seismic analysis. In fact, current methods of analysis provide important insights into the way that structures respond to earthquakes, and hence the ways in which designers can control this response. Moreover, a basic understanding of analytical principles is essential for enabling an informed and critical use to be made of computer-generated results, which currently form the basis for so much seismic analysis and design. Therefore in the authors' view, earthquake engineers must make the effort to understand the basics of dynamic seismic analysis. To gain a thorough understanding of the subject, reference needs to be made to some of the classic texts on the subject, for example Clough and Penzien (1993), which are however lengthy and require considerable effort.

This chapter does not attempt to reproduce this literature; its aim is to give the reader unfamiliar with the subject an outline of the most important principles and analytical methods and some help with jargon employed. It is worth repeating that despite its sophistication and fascination, analysis for seismic response gives results which are almost always beset with large uncertainties. The analysis is only a stage in the design process, and pages of computer output or complex mathematics should never be used as a replacement for sound engineering judgement.

3.2   Basic principles of seismic analysis

Seismic forces in a structure do not arise from externally applied loads. They are therefore different from more familiar effects such as wind loads, which are caused by external pressures and suctions on a structure. Instead, response is the result of cyclic motions at the base of the structure causing accelerations and hence inertial forces. The response is therefore essentially dynamic in nature and the dynamic properties of the structure, such as natural period and damping, are crucial in determining that response. Any seismic analysis, if it is to be at all realistic, must allow for this dynamic character, even if it is only in a simplified way.

The dynamic nature of the response is clearly a complicating factor, but there is a further analytical difficulty. As explained in Chapter 1, most engineered buildings are designed to withstand extreme earthquakes by yielding substantially. The designer must therefore have some understanding of the non-linear dynamic response of structures under extreme cyclic excitation. In principle, this poses very complex analytical problems. In practice, a combination of highly simplified analytical methods and appropriate design and detailing are often sufficient to secure satisfactory behaviour. However, it is essential to understand the basis and limitations of such techniques.

The rest of this section contains a series of 'snapshots' of the crucial issues, which it is hoped will give the reader some insight into the implications for design which are further discussed in subsequent chapters. The remaining sections then go on to give an outline of the main types of analysis currently in use.

3.2.1   Resonance

Almost everyone has experienced the phenomenon of resonance, for example the juddering that only occurs at a particular speed when driving a car with an unbalanced wheel. Resonance takes place when the period of excitation (in this example, the time for one revolution of the unbalanced wheel) matches the natural period of the structure.

Figure 3.1 shows the familiar curves for the steady-state response of a simple system subject to constant sinusoidal ground motion. The response is shown here in terms of peak acceleration of the system, divided by the peak ground acceleration to give a normalised response; it shows a marked peak when the system period matches that of the input motion, causing resonance.

By contrast, very rigid systems with low periods track the ground motion closely. The normalised response therefore tends to unity as the system period

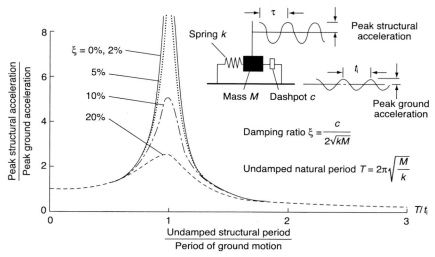

Fig. 3.1 Steady-state response to sinusoidal ground motion

tends to zero, or (equivalently) as the ground motion period becomes very long in comparison to the period of the structure.

Very flexible springs, on the other hand, act to isolate their masses from the input motion and so response tends to zero where the period of the structure is very long compared to that of the ground motion. In other words, response becomes small for very long-period structures or for very short-period motions. This is the principle behind, for example, isolation mounts for rotating machinery and also (as discussed later) seismic isolation systems for earthquake-resistant buildings.

Figure 3.1 describes the steady-state response to constant-amplitude single-period motions. By contrast, earthquakes are transient phenomena and the associated ground motions contain a range of periods. Nevertheless, certain periods tend to predominate, depending chiefly on the magnitude of the earthquake and the soil conditions at the site. The match between these predominant periods and the periods of a particular structure is crucial in determining its response. Figure 3.2 shows a response spectrum for a typical earthquake. Response spectra have already been introduced in subsection 2.5.2 and are discussed in more detail in subsection 3.2.5; the similarity in broad outline between Figs 3.1 and 3.2 is however immediately apparent. Thus, at zero period, the normalised response is unity. As the structural period increases, the trend (despite the spikiness for low levels of damping) is to increase to a maximum and then reduce to a level eventually approaching zero. Predominant ground motion periods at a firm soil or rock site are typically in the range 0.2–0.4 s while periods can reach 2 s or more on very soft ground. Since building structures have fundamental periods of approximately $0.1N$ (where N is the number of storeys), it can be seen that resonant amplification may well take place in common ranges of building height.

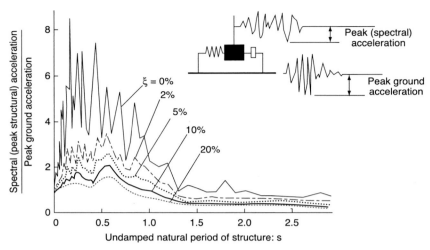

Fig. 3.2 Acceleration response spectrum for El Centro earthquake of 1940

3.2.2 Damping

When the cyclic excitation on a structure ceases, its response tends to die away. This is the phenomenon known as damping. Figures 3.1 and 3.2 show that the level of damping has an influence on response that may be as important as structural period.

If the damping is assumed to be 'viscous', i.e. the damping force varies with the velocity of the system relative to the ground, the mathematics become reasonably easy to solve. For this reason, the assumption of viscous damping is often adopted in analysis, although practical mechanisms of damping in buildings often follow somewhat different patterns, as discussed later. Viscous damping is usually expressed in terms of percentage of critical damping ξ, where $\xi = 100\%$ (critical damping) is the lowest level at which a system disturbed from rest returns to equilibrium without oscillation (Fig. 3.3). The percentage reduction between successive peaks in a cycle is approximately $2\pi\xi\%$, for small values of ξ; thus in

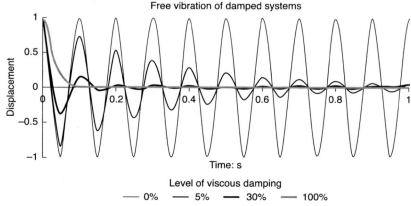

Fig. 3.3 Effect of viscous damping level on the decay of free vibrations

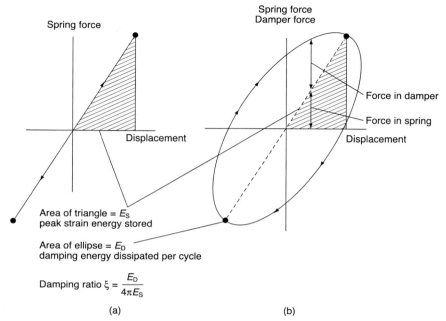

Fig. 3.4 Energy stored and dissipated in damped and undamped systems: (a) undamped system; and (b) damped system

Fig. 3.3, the 5% damped system reduces from an initial displacement of 1 to $(1 - 0.05 \times 2\pi)$ or about 0.7 after one cycle. Note that 5–7% damping represents the upper bound of damping found in most building structures responding at or around their yield point, while 30% damping represents an achievable level with the introduction of special engineered damping devices.

It can be shown that for a sinusoidal excitation, ξ is related to the ratio of energy dissipated by damping per cycle to the peak elastic strain energy stored (Fig. 3.4), a useful result for appreciating the physical significance of ξ.

A well-known text book result is that peak steady-state response at resonance under single-period excitation is approximately $(1/2\xi)$ times the input motion. Hence, the resonant response becomes infinite as the damping falls to zero. For the transient condition of an earthquake excitation, Fig. 3.2 shows a lower level of amplification at resonance; typical ratios of peak response to input are 2.5–3 for $\xi = 5\%$ (compared with 10 in Fig. 3.1 for constant sinusoidal excitation) and 5–8 for $\xi = 0\%$ (compared with infinity for Fig. 3.1). At or near resonance, therefore, earthquake response is less sensitive to damping level than steady-state sinusoidal response. However, Fig. 3.2 shows that response is more dependent on damping for earthquake excitations at periods away from resonance than is the case for single-period excitations.

Damping in buildings arises from a variety of causes, including aerodynamic drag (usually small), friction in connections and cladding (typically around 1% at amplitudes well below that corresponding to yield), damping associated with the soil and foundations (important in modes of vibration involving large soil deformation), and bond slip and cracking in reinforced concrete. These causes predominate when stresses are generally below yield. Plastic yielding gives rise to a different source of energy dissipation. Here, the damping energy is dissipated plastically as the structure cycles through hysteresis loops (Fig. 3.4), rather than as a result of viscous drag; hence the damping is referred to as 'hysteretic' rather than 'viscous'. An important difference between the two is that in hysteretic damping, the dissipated energy is proportional to peak displacement, while in viscous damping it is proportional to the displacement squared. Modelling hysteretic damping in plastic structures by an equivalent level of viscous damping therefore has some limitations which must be borne in mind.

3.2.3 Determining structural periods of buildings

As already discussed, the natural period and damping of a structure are the crucial parameters in determining its response to an earthquake ground motion. In the next two subsections, the determination of these two structural parameters is discussed.

The period of an undamped mass supported on a spring is equal to $2\pi\sqrt{(M/k)}$, as shown in Fig. 3.1. Doubling the mass therefore increases the period by about 40%, and the same is true if the stiffness is halved. More complex structures can have their natural periods determined from their mass and stiffness. A useful approximation for buildings with a regular distribution of mass and stiffness is

$$T \cong 2\sqrt{\delta}\,(\text{seconds}) \tag{3.1}$$

where δ is the lateral deflection in metres of the top of the building when subjected to its gravity loads acting horizontally; see for example Eurocode 8 Part 1 (CEN 2004) equation (4.9). Many structural analysis programs exist which produce more exact answers, although they are always worth checking by simple means, including those discussed in the next paragraph.

Theoretically-derived periods should always be treated with some caution. While the mass of a building structure may be reasonably easy to determine, its stiffness is usually much more uncertain. Non-structural elements such as cladding and partitions tend to add stiffness and hence to decrease natural periods. Moreover, the stiffness and hence the period depend on the amplitude of response primarily because when the structure starts to yield, the structural stiffness effectively decreases. An interesting paper by Ellis (1980) suggests that simple empirical formulae based on building height provide more accurate predictions of fundamental period than even quite sophisticated analyses. Most codes of practice including Eurocode 8 (CEN 2004) and IBC (ICC 2003) provide empirical formulae of this kind. Moreover, the IBC code requires that if the empirically-derived period results in substantially higher seismic forces than those corresponding to an analytically-derived period (i.e. brings the structure closer to resonance), then the forces based on the analytical period must be increased.

3.2.4 Determining damping level in buildings

Figures 3.1 and 3.2 show that the level of damping greatly influences response. Unlike period, damping can only be determined empirically and measurements in buildings show that the level varies over a large range in practice. It is found to be highly dependent on amplitude; for low to moderate levels of excitation (applicable to serviceability considerations) damping levels are generally in the range 1–2% of critical, while for levels of excitation with stresses approaching yield, damping may reach 3–10%. Concrete and masonry buildings tend to be at the higher end of the range, and steel at the lower end. Figures for design purposes are given in ASCE 4-98 (ASCE 1998).

A near-universal assumption is that damping in earthquake-excited buildings is 5%. Two provisos should be borne in mind when using this figure.

(1) It is only appropriate to severe earthquakes and would normally be unconservative for moderate events where yielding does not occur.

(2) This 5% damping represents the reduction in response associated with energy loss within the elastic range; the reduction is greatest at resonance, where it reduces the undamped response by a factor which depends on the particular earthquake motions but is typically about $2\frac{1}{2}$. Very much larger reductions are taken when allowing for post-elastic response, as described in subsection 3.2.9.

3.2.5 Earthquake response spectra

(a) General

Calculating the earthquake response, even of a simple structure idealised as a linear spring/mass/dashpot system (Fig. 3.2), is complex. Response spectrum analysis provides a much simpler method for calculating just the maximum response of the system during the earthquake, without having to calculate behaviour at other times. Since the maximum response is usually the quantity of greatest engineering interest, this is both useful and convenient. The method relies on the prior calculation of the maximum response of a series of simple systems with a range of periods from short to long and with various levels of damping. The maxima (called spectral values) are then plotted against the natural period of the system to produce the response spectrum shown in Fig. 3.2. Spectra can be plotted for spectral acceleration, velocity or displacement.

It is easy to show that the spectral (i.e. peak) response of all idealised linear systems with the same period and percentage of critical damping is the same for a given earthquake motion. Thus, a 10-tonne mass with 5% damping and 1-second period deflects and accelerates just as much as a 10 kg mass with the same damping and period when subjected to, say, the motions recorded during the El Centro earthquake of 1940. The response spectrum therefore becomes a powerful and versatile design tool. Knowing the mass, damping and period of a structure (providing it can be idealised as a simple linear spring/mass/dashpot system) and given an acceleration response spectrum, the following quantities of interest to the designer can be derived.

$$F = MS_a \tag{3.2}$$

where F is the peak spring force, M is the mass and S_a is the spectral acceleration.

$$S_d = F/k \tag{3.3}$$

where S_d is peak deflection, F is peak spring force and k is spring constant.

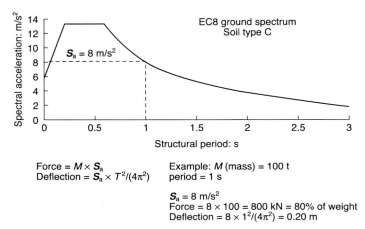

Fig. 3.5 Deriving peak force and deflection from a response spectrum

Combining equations (3.1), (3.2) and (3.3) gives

$$S_d = MS_a/(4\pi^2 M/T^2)$$

$$= S_a T^2/4\pi^2 \qquad\qquad (3.4)$$

Together, equations (3.2) and (3.4) show that with an earthquake response spectrum, two of the quantities of most use to earthquake engineers – namely peak force and peak deflection in a given earthquake – can be derived for a simple structure, provided its mass, natural period and damping are known. Figure 3.5 illustrates this with an example.

It should be noted that equation (3.2) relates to the peak spring force in a system, neglecting the damping force. Figure 3.4 shows that the total peak force due to spring and damper peaks just before the point of maximum displacement. Therefore S_a is slightly less than the true peak acceleration during an earthquake, and strictly speaking is defined as the 'pseudo-spectral acceleration' calculated from the peak deflection such that (by rearranging equation (3.4))

$$S_a = S_d(4\pi^2/T^2) \qquad\qquad (3.5)$$

Note that the quantity (mass times pseudo-spectral acceleration) represents the peak spring force within the system. Since the damping forces are in most cases fictitious quantities representing energy loss, it is the peak spring forces that are of most interest when assessing the structure's requirement for strength. In any case, for low levels of damping ($\xi < 20\%$) the difference between pseudo-spectral and true-spectral acceleration is very small. This is because the velocity is small at the time the true-spectral acceleration peak occurs, and so the damping force (which is proportional to velocity) is also small. Usually, therefore, acceleration response spectra refer to pseudo-spectral accelerations. Of course, for undamped systems, there is no difference between pseudo and true quantities.

(b) Smoothed design spectra
Each 'time history' of earthquake motions produces its own unique response spectrum, with a shape reflecting the frequency content of the motions. As explained in

Chapter 2, in design, a smoothed enveloped spectrum is used (Fig. 2.6), which irons out the spikes in response and effectively encompasses a range of different possible motions assessed for a particular site.

(c) Absolute and relative values

One common source of confusion in earthquake engineering relates to the fact that not only the structure, but also the ground, moves. Therefore, should motion be quoted relative to the ground or in absolute terms? It is particularly important to remember that spectral accelerations are always quoted as absolute values whereas spectral velocities and displacements are relative values, being the difference in motion between the mass and the ground. It may help to remember that the forces in the spring and dashpot result respectively from the relative deflection and velocity, but the absolute acceleration of the mass equals the spring plus dashpot force divided by the mass.

(d) Displacement spectra

All the response spectra shown so far have shown accelerations. These are of fundamental importance to the earthquake engineer because they relate to the maximum inertia (i.e. mass times acceleration) forces that develop during an earthquake and hence to the strength that a structure needs to resist those forces safely. However, spectra can also be drawn for peak displacement. It might be expected that there would be a close relationship between the displacement and acceleration spectra of a given earthquake, and equations (3.4) and (3.5) demonstrate that this is indeed the case. Given a displacement spectrum, a (pseudo-)acceleration can immediately be derived, and vice versa (Fig. 3.6); the acceleration spectrum equals the displacement spectrum times $4\pi^2$ divided by the period squared. It may therefore be wondered why it is necessary to have both types of spectra. The reason is that acceleration spectra are used for determining maximum strength requirements, while displacement spectra are used in simplified methods for

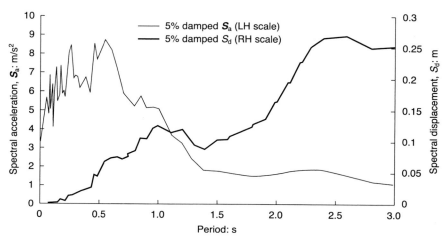

Fig. 3.6 Displacement and acceleration spectra

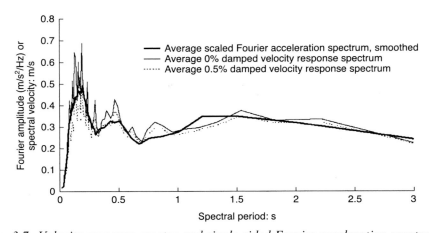

Fig. 3.7 Velocity response spectra and single-sided Fourier acceleration spectrum

assessing non-linear response in earthquakes, known as non-linear static analysis, as discussed further in subsection 3.4.3.

(e) Velocity and Fourier spectra

Velocity spectra can also be readily derived. For convenience, the 'pseudo-spectral' velocity is shown, which by analogy with equation (3.5) is defined as

$$S_v = S_d(2\pi/T) \tag{3.6}$$

This slightly overestimates the true peak velocity, although for low levels of damping, the discrepancy is small. Velocity is rarely a quantity of direct interest to earthquake engineers, and the primary importance of the velocity spectrum is that, for zero damping, it can be shown to be a fairly close upper bound to the single-sided Fourier acceleration spectrum of the relevant earthquake (Fig. 3.7). Fourier spectra can be used to derive power spectral densities (see for example ASCE 4-98 (ASCE 1998, equation 2-4.1)) which are used in probabilistic analysis methods and are also specified in ASCE 4-98 when checking the adequacy of time histories for design purposes.

(f) Capacity displacement spectra

Equation (3.4) shows that the three quantities – spectral acceleration, spectral displacement and structural period – are uniquely related for a specified level of damping; given two of them, the third is always known. So far, acceleration spectra (spectral acceleration plotted against structural period) and displacement spectra (spectral displacement against structural period) have been discussed. There is however a third possibility, namely plotting spectral acceleration directly against spectral displacement, and the result is called a capacity displacement spectrum (Fig. 3.8). The 'capacity' of the title refers to the fact that by multiplying the vertical (acceleration) axis by mass, a peak spring force is obtained, and the 'non-linear static' method of analysis relates this force to the capacity of a structure. Equation (3.4) can be rearranged to the form

$$T = 2\pi\sqrt{(S_d/S_a)} \tag{3.7}$$

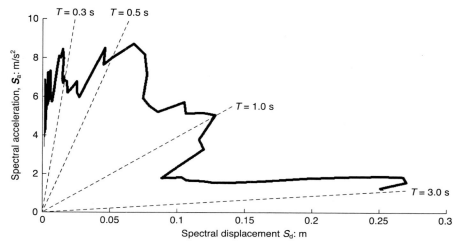

Fig. 3.8 Capacity displacement spectra

Constant ratios of S_d/S_a therefore represent constant values of structural period, as shown in Fig. 3.8. Discussion of the use of capacity spectra in engineering design and analysis is given in subsection 3.4.3.

3.2.6   Systems with multiple degrees of freedom (MDOF)

Almost all practical structures are much more complex than the 'single degree of freedom' (SDOF) spring/mass/dashpot systems shown in Figs 3.1 and 3.2, which have been discussed so far. However, many structures can be idealised as SDOF systems. A water tower with a rigid tank full of water supported by a relatively light frame is an example. The seismic response of many buildings is dominated by their fundamental sway mode and this fact can be used to create an SDOF idealisation. Irvine (1986), in his excellent text on dynamics, shows how many systems encountered in engineering practice can be treated in this way.

More complex structures need to be analysed by considering not only the fundamental mode but also the higher natural modes of vibration, which are a characteristic of the stiffness and mass distribution of the structure. These natural mode shapes, which are a structural property independent of the forcing vibration, are shown in Fig. 3.9 for a typical ten-storey building. Many computer programs exist to perform this calculation. A computer model of the structure must be established, using a combination of stick and shell elements as required with suitable support restraints, just as would be required for a normal static analysis. However, in addition the mass of the structure must be specified by adding mass elements to the model. With this mass and stiffness information, calculation of mode shapes and periods is a standard calculation performed by many structural analysis packages.

It transpires that the response of a linear structure can generally be calculated by considering the response in each of its modes separately (Fig. 3.10) and then combining the separate modal responses. This is possible because each mode of

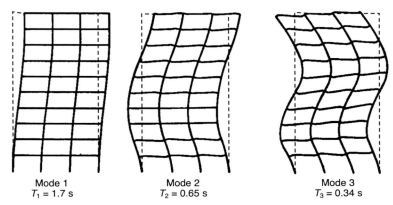

Mode 1   Mode 2   Mode 3
$T_1 = 1.7$ s $T_2 = 0.65$ s $T_3 = 0.34$ s

Fig. 3.9 Mode shapes and periods of a ten-storey frame building

vibration has an associated unique period and also a unique mode shape; therefore one parameter (e.g. the top deflection) is sufficient to define the entire deformation of the structure in that mode. In effect, therefore, each mode is an SDOF system. The basic form of equations (3.2) and (3.4) still holds, but the equations must be modified as follows. For the base shear in each mode, the total mass in equation (3.2) must be replaced by the appropriate 'effective' mass, which is always less than the total mass. For deflections and accelerations at any point in the system in each mode, the spectral values S_a and S_d must be multiplied by a structural constant and the value of the mode shape at the point under consideration. Clough and Penzien (1993, pp. 617ff) give the values for a distributed system as follows

$$\text{Base shear in mode } i = (L_i^2/M_i)S_{ai} \tag{3.8}$$

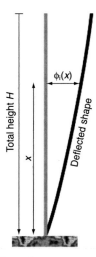

Fig. 3.10 Mode shape of a distributed system

Table 3.1 Modal multiplying factors for a uniform cantilever shear beam

Mode		1	2	3	4
Base shear factor $= \dfrac{(L_i^2/M_i)}{(\text{total mass})}$ Equation (3.8)		82%	8.0%	3.6%	1.2%
Acceleration factor $= (L_i/M_i)\phi_i(x)$ Equation (3.9)	Top	127%	40%	27%	16%
	Mid-height	90%	−28%	−19%	−11%
Deflection factor $= (L_i/M_i)\phi_i(x)$ Equation (3.10)	Top	127%	40%	27%	16%
	Mid-height	90%	−28%	−19%	−11%

Equation (3.8) is the equivalent for MDOF systems of equation (3.2), which applies to SDOF systems. As explained below, the term (L_i^2/M_i) has the dimensions of mass, and is always less than the total mass of the structure. A very useful result is that $\sum(L_i^2/M_i)$ summed over all modes is equal to the total mass. Therefore, sufficient modes must be considered in analysis to 'capture' an adequate proportion of the total mass. Codes often require that sufficiency is indicated when 90% of the mass is captured; for the example in Table 3.1, it can be seen that the first three modes capture 93.6% (i.e. 82% + 8% + 3.6%) of the total mass.

Similar equations apply for acceleration and displacement. These quantities obviously vary with height and so must include the mode shape, $\mu_i(x)$ (see Fig. 3.10).

$$\text{Acceleration at level } x \text{ in mode } i = S_{ai}[(L_i/M_i)\phi_i(x)] \qquad (3.9)$$

$$\text{Displacement at level } x \text{ in mode } i = S_{ai}[(L_i/M_i)\phi_i(x)](T^2/4\pi^2) \qquad (3.10)$$

where S_{ai} is the spectral acceleration corresponding to the ith mode period, $\phi_i(x)$ is the modal deflection at height x in mode i and L_i and M_i are structural properties defined in equations (3.11) and (3.12) below.

Equation (3.9) relates the peak acceleration at any level of the structure in a particular mode to that of its SDOF equivalent. The term $[(L_i/M_i)\phi_i(x)]$ is a dimensionless constant, which Table 3.1 shows can be either greater or less than 1. Thus, the acceleration at the top of a building swaying in its first mode is 27% greater than for its SDOF equivalent, but at mid-height it is 10% less.

Similar remarks apply to the peak relative displacement, which is given by equation (3.10), the MDOF equivalent of equation (3.4).

For distributed two-dimensional systems, L_i and M_i are calculated from equations (3.11) and (3.12).

$$L_i = \int_0^H m(x)[\phi_i(x)]\,\mathrm{d}x \qquad (3.11)$$

$$M_i = \int_0^H m(x)[\phi_i(x)]^2\,\mathrm{d}x \qquad (3.12)$$

where $m(x)$ is the mass per unit length at height x and $\phi_i(x)$ is the modal deflection at height x in mode i.

Since there are as many modes as degrees of freedom, changing to a modal analysis at first sight does not appear to help much. However, it transpires that

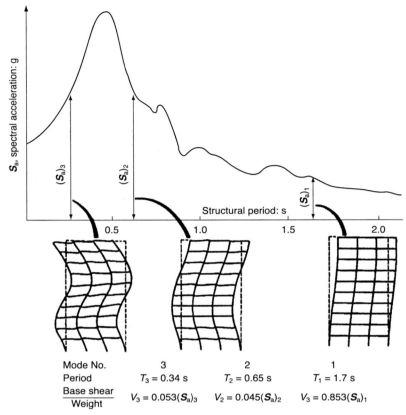

Fig. 3.11 Modal response spectrum analysis of an MDOF structure

the effective masses of the higher modes (i.e. the term (L_i^2/M_i) in equation (3.8) are low in the case of many practical structures. Therefore, a good approximation to response can usually be obtained from considering only the first few modes of vibration (and often only the lowest in each horizontal direction). For example, Table 3.1 shows the multiplying factors for a uniform cantilever shear beam such as shown in Fig. 3.10. The base shear in the first mode is 82% of that for a lumped mass/spring system with the same mass and period, while in the second mode the ratio drops to only 8% and 1.2% in the fourth mode. The acceleration and deflection at the top of the cantilever are 27% greater than for the equivalent SDOF system in the first mode, but substantially less in other modes.

The results of such an analysis give the maximum response of the structure for each mode of vibration. Although it is rigorously correct to add the response in each mode at any time to get the total response, the *maximum* modal responses, calculated from response spectrum analysis, do not occur simultaneously. Therefore a simple numerical addition of maximum modal responses usually results in a significant overestimate of the real maximum. The SRSS (square root of the sum of the squares) combination of modal responses (whereby each modal response is squared and the square root of the sum of all such squared response

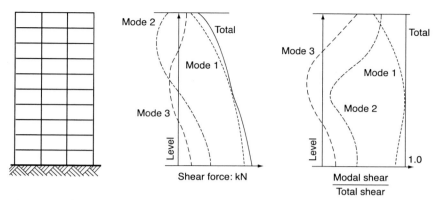

Fig. 3.12 Modal contributions to shear force in a typical frame building

is calculated) usually gives a good estimate of the true overall maximum, but it is only an estimate. Circumstances in which the SRSS estimate may be significantly unconservative are where there is significant response in two or more modes with very similar natural periods (the more sophisticated CQC (complete quadratic combination) method allows for this) and where there is significant response in modes with periods lower than the predominant periods of the earthquake motions (an effect not allowed for in CQC). In the latter case, a safe approximation is to add very short period responses; a less conservative method is given in section 3.7.2.1 of ASCE 4-98 (ASCE 1998). Most commercial computer programs that provide response spectrum analysis include SRSS and CQC combination methods; they are further discussed in ASCE 4-98.

The effect of these combination methods is that the fundamental mode is likely to contribute most of the base shear (unless of course the spectral accelerations of other modes are very much higher). This explains the previous assertion that, in many cases, a building can be treated as an SDOF system corresponding to the fundamental mode. However, where the first mode is well out of resonance with the earthquake motion but the second and third mode periods are close to resonance (a common situation for buildings of more than 20 storeys), shears and deflections at higher levels are likely to be strongly influenced by higher modes, as illustrated in Fig. 3.12. Hence there is usually a need to carry out a multi-mode analysis, rather than just a fundamental-mode analysis, for tall buildings.

There is an important consequence of the differing contributions of different modes to shear over the height of a structure, namely that the maximum shear force at any level is unlikely to occur simultaneously with the maxima at other levels or with maximum bending moments. Shear and bending moment diagrams obtained from a response spectrum analysis are therefore enveloped maxima and are not an equilibrium set of actions. In particular, maximum shear force does not equal rate of change of maximum bending moment, as would be the case in a conventional static analysis.

The preceding paragraphs have given a simplified account of multi-mode response spectrum analysis, which is the most common type of dynamic analysis currently performed in engineering design practice outside Japan. The advantages and disadvantages of the method are described further in subsection 3.3.2.

3.2.7 Torsional response

Seismic ground motions are predominantly translational, not rotational. However, where the centres of mass and stiffness of a structure do not coincide, coupled lateral–torsional response occurs (Fig. 3.13). Structures with significant torsional eccentricity are found to have a much worse performance during earthquakes. Coupled lateral–torsional response cannot of course be analysed using two-dimensional models, since three-dimensional behaviour is involved. Static analysis by applying code-required forces at the centre of mass may underestimate response because of dynamic effects, as discussed by Chandler (1990); a possible example is where the period of the lateral–torsional mode of vibration matches the predominant period of the earthquake. Linear dynamics may underestimate the response after yielding, because the less stiff side tends to yield first, becoming even more flexible and hence adding to the eccentricity (Bruneau and Martin 1990).

Codes of practice treat torsion in the following ways.

- Requiring additional strength (up to 50% in Japanese codes and 20% in Eurocode 8) beyond that indicated by analysis.
- Requiring more sophisticated analysis if eccentricity exceeds prescribed limits. For example, explicit non-linear dynamic analysis is specified by Japanese codes and three-dimensional modal response spectrum analysis by Eurocode 8 and US codes.
- Where static analysis is permitted, the IBC code requires the distance between point of application of lateral load and the centre of stiffness to be increased, if the eccentricity exceeds a threshold.
- Codes usually specify an 'accidental' torsion; that is, an offset of point of application of lateral load by 5–10% of building dimension from centre of mass.

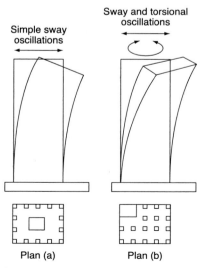

Fig. 3.13   Couple lateral–torsional response

3.2.8 P–delta effects

Lateral deflections give rise to gravity-induced moments (Fig. 3.14). Usually the moments are small, but where the product of gravity load (**P**) and the lateral deflection (δ) is a significant fraction of the seismic overturning moment, the resulting '**P**–delta' effect should be allowed for. It can be easily incorporated into a non-linear analysis, but needs special techniques to include in a linear-elastic analysis; most standard linear-analysis computer programs do not allow for it.

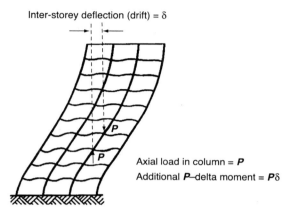

Inter-storey deflection (drift) = δ

Axial load in column = **P**
Additional **P**–delta moment = **P**δ

Fig. 3.14 P–delta moments

Codes such as Eurocode 8 and IBC state that **P**–delta effects can be neglected if specified deflection limits are not exceeded.

3.2.9 Non-linear response

(a) Ductility

Discussion has so far been in terms of linear-elastic response. However, most structures are designed to yield in the event of an earthquake and so post-yield response is often of crucial importance. As a result, the ductility (or lack of it) that a structure possesses becomes a vital consideration and so before discussing the non-linear effects involved, a definition of 'ductility' is required. Ductility is the ability of a structure to withstand repeated cycles into the post-elastic range without significant loss of strength. It can be quantified in terms of degree of plastic deformation. Figure 3.15 defines 'deflection ductility' for a simple yielding system. It is useful also to define the local degree of plastic deformation in terms of 'curvature ductility', the ratio of maximum curvature of the beam to curvature at first yield of the flexural steel. Suppose Fig. 3.15 represents deflection of the top of a building as a function of base shear. Figure 3.16 shows that some parts of the structure are likely to start to yield well before the nominal yield deflection D_{yield} is reached. Moreover, after the onset of yielding, further deformation tends to concentrate in the yielding regions, rather than in the parts of the structure that remain elastic. Therefore, the curvature ductility of the yielding beams will be many times the overall displacement ductility.

Figure 3.17 shows the same information in a different way. The available ductility μ based on the deflections of lower floors is likely to appear to be much greater

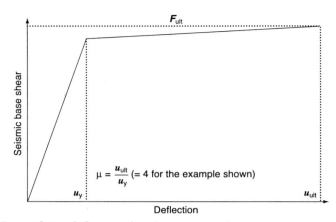

Fig. 3.15 *Quantifying deflection ductility in a simple system*

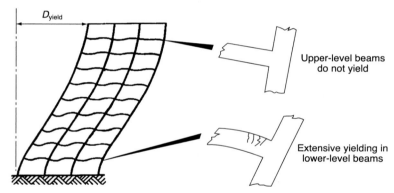

Fig. 3.16 *Relationship between curvature and deflection ductility in a building frame*

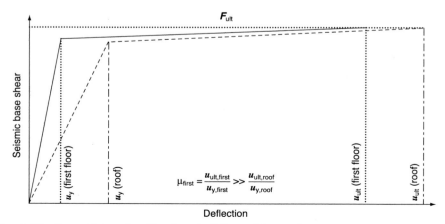

Fig. 3.17 *Quantifying ductility in a building with a soft storey*

than that based on the deflections of the roof. Therefore, curvature ductility relates more closely to ultimate deflection capacity than does displacement ductility.

(b) Ductility demand and supply

For the engineering designer, it is helpful to distinguish between ductility demand and supply.

The *ductility demand* an earthquake makes on a structure is defined as the maximum ductility that the structure experiences during that earthquake. Ductility demand is a function of both the structure and the earthquake; thus, in general, the demand decreases as the yield strength of the structure increases, and the demand increases as the intensity of the motions increases.

The *ductility supply* is, by contrast, a property only of the structure; it is defined as the maximum ductility a structure can sustain without fracture or other unacceptable consequences. Note that μ in Figs 3.15 and 3.17 quantifies ductility supply, not ductility demand.

The objective of the designer, therefore, is to ensure that ductility supply exceeds demand by a sufficient margin to ensure safe performance in the design earthquake. A major purpose of seismic analysis is to establish the level of ductility demand in a structure; the equally important design measures to ensure the existence of an adequate ductility supply are at the heart of most current seismic codes of practice, as discussed in later chapters.

(c) Ductility-modified spectra for SDOF systems

Just as a linear response spectrum gives the maximum response of a linear SDOF system to a given earthquake, so ductility-modified spectra can be developed for the response of a ductile SDOF freedom. In constructing ductility-modified spectra, the yield strength of the SDOF is chosen so that the ductility demand during a given earthquake is limited to a given value μ. Families of ductility-modified curves can then be drawn corresponding to different global displacement ductility demands μ (see Fig. 3.18). For each value of μ, the yield strength has been set such that the peak displacement equals μ times the yield displacement.

Figure 3.18 is the non-linear equivalent of Fig. 3.2, differing only in showing the response for different maximum ductility demands rather than different levels of

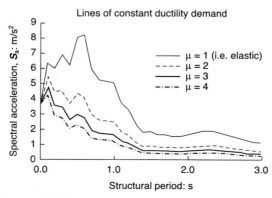

Fig. 3.18 Ductility-modified response spectra

hysteretic damping. The similarity between the two figures is not a coincidence; the increasing ductility demands in Fig. 3.18 represent increasing amounts of damping, although hysteretic damping is involved, rather than the viscous damping of Fig. 3.2. Note particularly that high ductility demands are ineffective in reducing response in very stiff structures with structural periods close to zero, in just the same way that applies to viscous damping. This is because both viscous and hysteretic damping arise from internal structural deformations (represented as compression of the spring in an SDOF idealisation). These structural deformations are relatively small compared with the ground movements in very stiff structures, and hence give rise to relatively small reductions in response.

For an elastic-perfectly plastic SDOF system, equation (3.4) becomes modified as follows

$$S_{d\mu} = \mu \frac{F_y}{k} = \mu m S_{a\mu} \times \frac{T^2}{4\pi^2 m} = \mu \frac{S_{a\mu} T^2}{4\pi^2} \tag{3.13}$$

Here, k is the pre-yield stiffness of the spring, and F_y is the yield force in the spring. T is the period of the structure before it yields. The logic behind equation (3.13) is as follows. The deformation at yield is F_y/k, and (by definition) the maximum deformation $S_{d\mu}$ under the earthquake is $\mu F_y/k$, since the yield strength of the structure has been set to achieve a global ductility demand μ. Equation (3.13) then follows directly. Note that it is exact (unlike the relationships shown in Fig. 3.19 below). The important implication of equation (3.13) is that in ductile structures, displacements are μ times greater than their elastic equivalents with the same level of stress, the reason being that plastic strains increase the displacements in the yielding structure.

Construction of ductility-modified spectra directly from the earthquake record is in principle straightforward using appropriate software, and a number of programs exist to do this. However, an approximate ductility-modified spectrum can be estimated much more directly from the elastic spectrum, as is now explained.

For structures with very long initial periods (i.e. very flexible structures), whether ductile or elastic, the maximum displacement equals the peak ground displacement; essentially the structure is so floppy that the structure stays where it is and the ground moves beneath it. Therefore, displacements of very flexible ductile and elastic structures are equal. The same result – equality of displacements in elastic and yielding structures – holds approximately for all structures where the initial period is greater than the predominant period of the ground motions (approximately 0.1–0.3 s for firm ground sites, 1 s or more for very soft sites).

Therefore, for medium to long period structures, it is quite easy to see that the acceleration in the plastic structure is a factor μ lower than its elastic equivalent. This is because the plastic structure suffers a force μ times lower than it would have done if it had remained elastic, but it experiences the same maximum deformation u_{ult} (see Fig. 3.19). However, it will almost certainly undergo some permanent deformation, and possibly 'damage' as well.

This result does *not* apply to stiff structures with periods lower than the predominant ones of the earthquake motion. For rigid structures ($T = 0$), the force

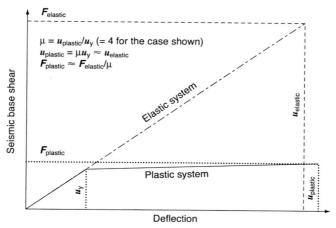

Fig. 3.19 *Forces and deflections in plastic and elastic systems: flexible structures (deflections preserved)*

in both yielding and elastic systems must be the same, and equal to the structural mass times peak ground acceleration. It follows that the deformation in the plastic (yielding) structure is μ times *greater* than its elastic equivalent (Fig. 3.20). Therefore, ductility (hysteretic damping) is no advantage to a very rigid structure, in just the same way as viscous damping (Fig. 3.2) does not reduce the response of rigid systems.

These results are summarised in Table 3.2 and can be used to construct an approximate ductility-modified acceleration response spectrum relatively easily

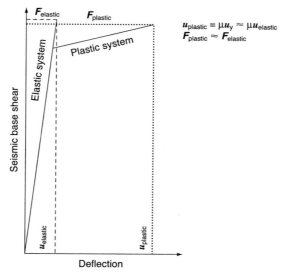

Fig. 3.20 *Forces and deflections in plastic and elastic systems: rigid structures (forces preserved)*

Table 3.2 Comparison of forces and deflections in elastic and yielding (plastic) structures

	Elastic structure	Yielding structure
Medium to long period structures:		
Acceleration	$S_{a_{elastic}}$	$S_{a_{plastic}} \approx S_{a_{elastic}}/\mu$
Force	$F_{elastic}$	$F_{plastic} \approx F_{elastic}/\mu$
Deformation	$u_{elastic}$	$u_{plastic} \approx u_{elastic}$
Very short period structures:		
Acceleration	$S_{a_{elastic}}$	$S_{a_{plastic}} \approx S_{a_{elastic}}$
Force	$F_{elastic}$	$F_{plastic} \approx F_{elastic}$
Deformation	$u_{elastic}$	$u_{plastic} \approx \mu u_{elastic}$
Short to medium period structures:		
Acceleration	$S_{a_{elastic}}$	$S_{a_{plastic}} \approx S_{a_{elastic}}/X$
Force	$F_{elastic}$	$F_{plastic} \approx F_{elastic}/X$
Deformation	$u_{elastic}$	$u_{plastic} \approx \mu/X\, u_{elastic}$
where X is a factor between 1 and μ		

from the corresponding elastic spectrum. For periods above the spectral peak, the ductility-modified spectral acceleration $S_{a\mu}$ is obtained by dividing the elastic spectral acceleration S_a by μ, the global displacement ductility factor. For rigid systems, the ductility-modified and elastic spectral accelerations are equal. For intermediate periods, the elastic spectral acceleration is divided by a factor between μ and 1. In Eurocode 8, the reduction factor increases linearly between 1 and μ as the structural period increases from 0 s to the period at the start of the spectral peak value (between 0.05 and 0.20 s, depending on the type of soil and magnitude of earthquake). Other more complex relationships have been published, but the one in Eurocode 8 should be sufficient for most purposes, given the other uncertainties involved.

(d) Application of ductility-modified spectra to MDOF systems

Subsection 3.2.6 showed how a rigorously correct analysis of a linear elastic MDOF (multiple degrees of freedom) system was possible on the basis of a response spectrum constructed for a linear SDOF (single degree of freedom) system. It might be thought that a similar extension of a ductility-modified response spectrum from SDOF to MDOF would be possible. In other words, an MDOF structure, for which the ductility demand was required to be μ, could be analysed for accelerations and forces as if it were an elastic structure, but with the elastic ground acceleration spectrum replaced by a ductility-modified spectrum corresponding to the required value of μ. The yield strengths necessary to limit the ductility demand to μ would be obtained directly from such an analysis, but (from equation (3.13)), all the displacements would need to be increased by a factor μ. The assumption that the analysis of a ductile MDOF is possible in this way underpins most practical and code-based designs using response spectrum analysis.

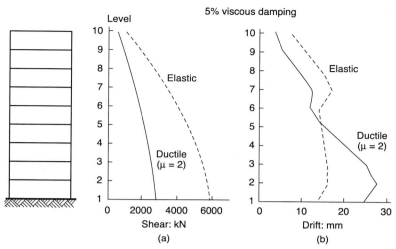

Fig. 3.21 Comparison of forces and deflections in the yielding and elastic response of a typical ten-storey building: (a) forces; and (b) deflections

Unfortunately, there is no fundamental reason why such an analysis should apply. As explained in subsection 3.2.6, the linear MDOF analysis works by effectively splitting the structure into a series of SDOF systems, each representing one of its natural modes. However, once the structure yields, the unique mode shapes on which the linear analysis relies no longer apply, and the modal periods start to increase.

Where yielding is spread uniformly through the structure, and the deformed shape is similar to the elastic first mode shape, then ductility-modified response spectrum analysis may give reasonable answers, but it must be remembered that the answers from such analysis are never exact. Figure 3.21 compares two analyses of the same structure subjected to the same ground motion. In the first analysis, the structure remains elastic and experiences a base shear of 6 MN. In the second analysis, which used rigorous non-linear time-history methods, and not a ductility-modified spectrum, the structure had the same initial stiffness but was allowed to develop a displacement ductility μ of 2.

As expected, the base shear in the yielding structure has halved to just under 3 MN, with similar reductions (i.e. a reduction factor of $1/\mu$) applying throughout the height of the building. However, the storey drifts (difference in deflections between one storey and the next) which ductility-modified response spectrum analysis would have predicted to be the same as in the elastic case, in fact were very different: the drifts at the bottom were about double the elastic values, and at the top about half – a fairly typical result for a structure (such as this) where yielding is well distributed throughout the structure.

The results of ductility-modified response spectrum analysis will be much more in error for structures where the yielding is concentrated at one level – for example, in 'soft storey' structures. As discussed later, non-linear time-history analysis addresses these problems in the most complete way, while displacement-based design, using non-linear static (pushover) methods of analysis, also accounts for this non-uniform distribution in displacements in a less complex way.

3.2.10   Consequences of yielding response

There are a number of important consequences of yielding which a designer should bear in mind.

(1) Member forces remain well below the level they would have reached, had the structures remained elastic (Fig. 3.21(a)). The reduced response is due to the hysteretic damping associated with the yielding. For structures with an initial period greater than the predominant period of the earthquake, the lengthening of the structural period caused by yielding will also help to reduce response.

(2) Post-yield deformed shape is markedly different from the elastic condition (Fig. 3.21(b)). In the yielding areas (usually the lower levels of a building), deformations tend to be greater than elastic values, while in other areas they tend to be less. Therefore, the implicit assumption in most codes that the deflections remain equal to the elastic deflections corresponding to 5% damping is likely to be unconservative at the lower levels of buildings.

(3) Members are damaged; this can be thought of as a low-cycle fatigue effect. Hence the number as well as the magnitude of the yielding cycles is important.

(4) In redundant (hyperstatic) structures such as frames, gravity moments become significantly redistributed, which may significantly affect the frame's earthquake-resisting properties.

(5) The increase in ratio of deflection to restoring forces means that P–delta effects (Fig. 3.14) become relatively more important.

These five effects can be dealt with by using various approaches.

(1) The member force reduction due to ductility is allowed for (at least in part) by reducing elastic forces by factors such as R in IBC or q in Eurocode 8, which depend on the available ductility. Eurocode 8 (unlike IBC) takes full account of the reduced effectiveness of ductility in very stiff structures.

(2) Post-yield deflections calculated from a ductility-modified response spectrum analysis should be treated with caution, for the reasons discussed above.

(3) Low-cycle fatigue effects are generally dealt with by appropriate detailing rather than direct analysis. For example, code rules for provision of transverse steel at a potential plastic hinge location of a reinforced concrete beam are greatly influenced by the need to prevent the flexural stiffness and strength from degrading during repeated cycles of yielding.

(4) Moment redistribution can have significant effects in frames where gravity loading produces moments which are a substantial fraction of the yield moments (Fenwick *et al.*, 1992). However, most codes either do not consider the effects of moment redistribution (e.g. IBC) or only partially account for it (e.g. New Zealand code NZS 3101).

(5) **P**–delta effects are rarely required to be considered by codes, as discussed in subsection 3.2.8 above, but this may be unconservative (Fenwick and Davidson 1987).

3.2.11 Other important considerations for a seismic analysis

(a) Influence of non-structure

Non-structural elements such as cladding and partitions are not usually explicitly allowed for in analysis but they may have an important and not always beneficial effect on response. For example, cladding can stiffen a structure and may bring its natural period closer to resonance with the predominant period of an earthquake.

As another example, infill blockwork which is not full height may create a short column whose shear strength is less than its bending strength and which is therefore prone to brittle failure (Fig. 1.17).

The designer has two alternatives. Either non-structural elements can be fully separated from the main structure, or the interaction between structure and non-structure must be allowed for in analysis. The first alternative creates a more predictable system, but may well lead to its own problems; for example, separation joints between infill masonry and structural frames are hard to detail satisfactorily to provide the weatherproofing and out-of-plane restraint that is needed. The second alternative can sometimes lead to satisfactory results; for example, the increased strength and stiffness provided to a structural frame by rigid infill masonry panels may more than offset the reduction in ductility and predictability.

(b) Site-effects

The nature of the soils at a site can have a dominating influence on the seismic motions at the site (Figs 2.7 and 2.8) and may also significantly affect the dynamic characteristics of structures built there, by increasing the foundation flexibility. These important considerations are discussed in Chapter 4.

3.3 Linear elastic forms of seismic analysis

Various forms of linear and non-linear analysis are possible, which build on the theoretical basis set out in section 3.2. This section describes the main linear analysis methods permitted in codes of practice, while the next section describes the non-linear methods.

Where significant ductility is assumed in design, a structure designed purely on the basis of a linear elastic analysis may well be unsafe. Where no explicit non-linear analysis is performed, minimum provisions are essential to ensure satisfactory post-yield behaviour, as discussed further in section 3.5.

3.3.1 Equivalent linear static analysis

All *design* against earthquake effects must consider the dynamic nature of the load. However, for simple regular structures, *analysis* by equivalent linear static methods is often sufficient. This is permitted in most codes of practice for regular, low- to

Table 3.3 Parameters to consider in a simple seismic analysis

Parameter	Symbol in Eurocode 8 (CEN 2004)	Symbol in IBC (ICC 2003)
Geographical location	a_{gR}, design ground acceleration on rock or firm ground (See also note 1 below)	S_s and S_1 (spectral accelerations at short period and at 1 s)
Foundation soils	S, soil parameter	Site class
Intended use, which influences acceptable level of damage	γ_1, importance factor	I_E, seismic importance factor
Structural form, which influences the available ductility	q, behaviour factor	R, response modification factor
Weight of structure and contents	$\sum G_{kj} + \sum \psi_{EI} Q_{ki}$, full characteristic dead load plus reduced characteristic live load	W, effective seismic weight (full dead load plus reduced live load)
First-mode period of the structure	T	T

Note 1: In Eurocode 8, geographical location also determines the choice of Type 1 or 2 spectral shapes, which accounts for whether sites are influenced by earthquakes of larger (Type 1) or smaller (Type 2) magnitudes.
Note 2: In both Eurocode 8 and IBC, structural irregularities in plan and elevation may lead to increased strength requirements.

medium-rise buildings and begins with an estimate of peak earthquake load calculated as a function of the parameters shown in Table 3.3. However, the following should be borne in mind.

For example, the seismic base shear F_b in Eurocode 8 Part 1, Section 4 is given by

$$F_b = S_d(T)\left(\sum G_{kj} + \sum \psi_{EI} Q_{ki}\right)\lambda \tag{3.14}$$

where $S_d(T)$ is the ductility-modified spectral acceleration for a period T, peak ground acceleration $a_{gR}\gamma_1$, behaviour factor q, and the appropriate soil type and spectral shape type. For example, the plateau value of $S_d(T)$ equals $2.5a_{gR}\gamma_1 S/q$.

λ = 0.85 for shorter period structures (around $T \leq 1$ s, depending on soil type)
= 1 for longer period structures.

λ = 0.85 for shorter periods corresponds to the modal reduction factor for base shear shown in Table 3.1. It is increased to 1 in tall buildings to allow for their greater potential importance and the increased influence of higher mode effects.

$(\sum \boldsymbol{G}_{kj} + \sum \psi_{EI} \boldsymbol{Q}_{ki})$ = mass considered for seismic loading
= full dead loads $\sum \boldsymbol{G}_{kj}$, plus live loads $\sum \boldsymbol{Q}_{ki}$ reduced by a reduction factor ψ_{EI} typically equal to 0.3.

The calculated load is then applied to the structure as a set of static horizontal loads with a prescribed vertical distribution, approximating to the first-mode response of a regular building.

The theoretical basis for equivalent static analysis is that the static forces are chosen to produce the same extreme deflected shape as would actually occur (momentarily) during the earthquake. For a structure responding in only one mode, the velocity is zero at all points in the structure when this maximum deflection is experienced. The equivalent static force therefore equals mass times acceleration (Dalambert force) at each point. Hence an exact equivalence between equivalent static and dynamic analysis is possible. However, where more than one mode is involved, different levels in the structure reach their extreme response at different moments of time and a single set of static forces can never truly represent the dynamic maxima at all levels. Equivalent static analysis can, therefore, work well for low- to medium-rise buildings without significant coupled lateral–torsional modes, in which only the first mode in each direction is of significance. Tall buildings (over, say, 75 m), where second and higher modes can be important, or buildings with torsional effects, are much less suitable for the method, and both Eurocode 8 and IBC require more complex methods to be used in these circumstances. However, it may still be useful, even here, as a 'sanity check' on later results using more sophisticated techniques.

3.3.2 Modal response spectrum analysis

With the advent of powerful desktop computers, this type of analysis has become the norm. It involves calculating the principal elastic modes of vibration of a structure. The maximum responses in each mode are then calculated from a response spectrum and these are summed by appropriate methods to produce the overall maximum response. The method was outlined in subsections 3.2.5 and 3.2.6.

The major advantages of modal response spectrum analysis (RSA), compared with the more complex time-history analysis described later, are as follows.

(1) The size of the problem is reduced to finding only the maximum response of a limited number of modes of the structure, rather than calculating the entire time history of responses during the earthquake. This makes the problem much more tractable in terms both of processing time and (equally significant) size of computer output.

(2) Examination of the mode shapes and periods of a structure gives the designer a good feel for its dynamic response.

(3) The use of smoothed envelope spectra (Fig. 2.6) makes the analysis independent of the characteristics of a particular earthquake record.

(4) RSA can very often be useful as a preliminary analysis, to check the reasonableness of results produced by linear and non-linear time-history analyses.

Offsetting these advantages are the following limitations.

(1) RSA is essentially linear and can make only approximate allowance for non-linear behaviour.
(2) The results are in terms of peak response only, with a loss of information on frequency content, phase and number of damaging cycles, which have important consequences for low-cycle fatigue effects. Moreover, the peak responses do not generally occur simultaneously; for example, the maximum axial force in a column at mid-height of a moment-resisting frame is likely to be dominated by the first mode, while its bending moment and shear may be more influenced by higher modes and hence may peak at different times.
(3) It will also be recalled (subsection 3.2.6) that the global bending moments calculated by RSA are envelopes of maxima not occurring simultaneously and are not in equilibrium with the global shear force envelope.
(4) Variations of damping levels in the system (for example, between the structure and the supporting soils) can only be included approximately. ASCE 4-98 (ASCE 1998) section 3.1.5 discusses ways of achieving this.
(5) Modal analysis as a method begins to break down for damping ratios exceeding about 0.2, because the individual modes no longer act independently (Gupta 1990).
(6) The method assumes that all grounded parts of the structure have the same input motion. This may not be true for extended systems, such as long pipe runs or long-span bridges. Der Kiureghian *et al.* (1997) have proposed ways of overcoming this limitation.

3.3.3   Linear time-history analysis

The complete 'time history' of response to an earthquake can be obtained by calculating the response at successive discrete times, with the time step (interval between calculation times) sufficiently short to allow extrapolation from one calculation time to the next. Where a linear analysis is involved, the time step should not exceed a quarter of the period of the highest structural mode of interest. This solution method in the 'time domain' is further discussed by Clough and Penzien (1993).

A linear time-history analysis of this type overcomes all the disadvantages of RSA, provided non-linear behaviour is not involved. The method involves significantly greater computational effort than the corresponding RSA and at least three representative earthquake motions must be considered to allow for the uncertainty in precise frequency content of the design motions at a site. With current computing power and software, the task of performing the number crunching and then handling the large amount of data produced has become a non-specialist task. More problematic is the choice of suitable input time histories to represent the ground motions at a site, as discussed in subsection 2.8.3.

3.3.4   Linear time-history analysis in the frequency domain

Linear time-history analysis can also be performed in the 'frequency domain', whereby the input motion is split into its single period harmonic components –

Fourier spectrum – by means of Fourier analysis. The analysis is performed by summing the separate responses to these harmonic components; it therefore can only be used for linear responses, where superposition is valid. The output is also obtained as a set of Fourier spectra, which can then be used to compute time histories of results in the time domain. The details and theoretical basis of the technique are described by Clough and Penzien (1993).

The possibility of increased computational efficiency when using frequency domain analysis is of less importance now, because of the ready availability of computing power. It is however sometimes used in soil–structure interaction analyses, since the flexibility of supporting soils can best be represented by frequency-dependent springs and this requires a frequency domain analysis. It is also the basis of some probabilistic methods, which have a wider application. As noted above, non-linear analysis is not possible in the frequency domain.

3.4 Non-linear analysis

3.4.1 Introduction

By its nature, linear analysis can give no information on the distribution of post-yield strains within a structure, and only limited information on the magnitude of any post-yield strains that might develop. The best that can be hoped for is that by means of an elastic analysis, the structural strength can be set to a level which will limit post-yield strains to acceptable levels. However, most structural failures during earthquakes occur as a result of elements experiencing strains beyond the limit that they can sustain. Non-linear analysis offers the possibility of calculating post-elastic strains directly, which is an enormous potential advantage. With the availability of increased computing power and more sophisticated software, non-linear methods are being increasingly used in design practice. It is noteworthy that simple non-linear time-history analyses were effectively mandatory for the seismic design of tall buildings in Japan since at least the 1980s.

3.4.2 Non-linear time-history analysis

Non-linear effects can be allowed for by stepping through an earthquake and extrapolating between calculation times, in just the same way as for a linear time-history analysis. The simplest (and most tractable) analytical models consist of frame elements in which non-linear response is assumed to be concentrated in plastic hinge regions at their ends. More sophisticated models can involve non-linear plate and shell elements.

Non-linear methods enable the most complete allowance to be made for the combination of dynamic response with the onset of plasticity and variation in time-dependent parameters such as the possible loss of strength and stiffness of plastic hinge regions under repeated large cyclical strains or the increase in pore water pressures in soils. Naturally, this extra information is bought with very considerably increased computational effort; the time steps used must be much less than those in a linear analysis. Clough and Penzien (1993) describe the solution techniques involved.

There are now a number of reasonably user-friendly commercial packages available which will carry out a non-linear time-history analysis and can analyse a practical size of building frame using a desktop computer. This is different from the situation of only 15 years ago, when non-linear time-history analysis was a difficult and time-consuming exercise; believable results are now reasonably easily obtained. However, there are still many pitfalls; results may be critically dependent on small variations in input parameters and sensitivity studies are likely to be needed, particularly since the non-linear cyclic response characteristics assumed in the computer model are probably only a crude and uncertain approximation of reality. At the very least, a range of different input motions must be used, since substantially different responses can be obtained from input motions with similar response spectra. Eurocode 8 requires at least three different time histories to be used, and this should be regarded as a bare minimum.

At the time of writing (2005), non-linear time-history analysis is still seen as a non-routine technique in design practice, needed only for special cases such as unusually important buildings or those with novel means of earthquake protection. However, the static non-linear analyses described below are become increasingly favoured.

3.4.3   Non-linear static and 'displacement-based' methods

(a) General

Non-linear static methods have recently gained wide currency, and offer the advantage of giving direct information on the magnitude and distribution of plastic strains within a structure, based on the ground motions represented by the design response spectrum in a code of practice, without the difficulties inherent in a non-linear time-history analysis, and the associated requirement to choose suitable ground motion time histories. As explained in more detail below, the method involves modelling a frame structure as an equivalent SDOF structure, whose properties have been determined by means of a 'static pushover' analysis performed on a non-linear model of the frame. The peak displacement of this SDOF structure is then determined directly from the design response spectrum, and then imposed on the frame model to determine the peak plastic strains in the frame, and their distribution.

Modelling a complex non-linear frame as an SDOF is clearly a drastic simplification, and the results can never be as 'accurate' as those obtained from more complex methods. In some cases, the method may be unsuitable; in particular, this is likely to apply if the building structure in question is subject to significant torsional response, since an SDOF idealisation can only capture translational and not torsional response. However, they address the huge drawback of the linear methods of analysis that have been standard in Western design practice, namely that such methods cannot properly capture the non-linear behaviour which characterises the intended response of most buildings during their design earthquake.

Non-linear static methods are often referred to as 'displacement-based' (as opposed to strength-based) design methods, since peak displacements, rather than peak strength, are more obvious during the process. However, the distinction

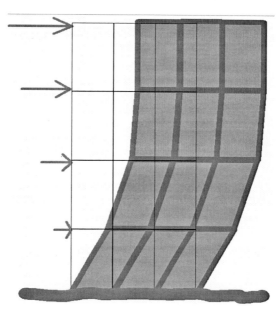

Fig. 3.22 Static pushover analysis

is somewhat artificial, since displacement and forces are inextricably linked in any method of analysis. Moreover, the phrase 'displacement-based design' usually relates to a method of analysis rather design.

The description of the methods that follows is based on US practice, for example as contained in FEMA 356 (FEMA 2000), although it has gained wider international acceptance. Annex B of Eurocode 8 Part 1 sets out a non-linear static method based on the 'N2' procedure described by Fajfar (2000). Eurocode 8 Part 2 for bridges and Part 3 for the assessment and retrofit of existing buildings also provide advice on non-linear static analysis.

A further source of information on non-linear analysis procedures is ATC-55 from FEMA 440 (FEMA 2005).

(b) Static pushover analysis
The first stage in the process is to perform a 'static pushover analysis' (Fig. 3.22). This involves defining a set of lateral forces, with a vertical distribution corresponding to those of the inertia forces developed in an earthquake, which are applied as a static loadcase to a non-linear model of the structure. All the forces are gradually increased by the same proportion, and the deflection of the top structure is plotted against the total applied shear; this is the basic pushover curve (Fig. 3.23). As yielding occurs in the structure, its properties are appropriately modified; for example, plastic hinges are introduced at the ends of yielding beams.

(c) Target displacement method
Having calculated a pushover curve, two methods are available to calculate the maximum deflection at the top of the building under the design earthquake

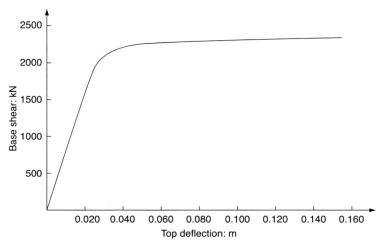

Fig. 3.23 Static pushover curve

motions. The most straightforward method to use is the target displacement method of FEMA 356; the methods set out in the various parts of Eurocode 8 are similar. Here, equation (3.4) is used to find the peak deflection of a linear SDOF with a period corresponding to the first mode of the building, using an elastic, 5% damped response spectrum. This is of course the 5% spectral deflection at the first mode period. If the structure remained elastic, if structural damping were 5% and if the first mode dominated the response, then the top deflection would equal this deflection increased by the modal factor, which from Table 3.1 is 1.27 for a uniform cantilever. For other than short-period structures, this elastic estimate often provides a good approximation for a plastically responding structure, even when the structure yields significantly (see Fig. 3.19). However, it is an underestimate for short-period structures (Fig. 3.20), or for structures subject to strength or stiffness degradation. FEMA 356 provides a detailed formula to relate the actual top deflection to the spectral deflection, based on these principles, which take account of building period, height, hysteresis characteristics etc.

Having calculated the top deflection of the building, the static pushover analysis is used to calculate the forces and plastic strains throughout the structure which correspond to this top deflection.

(d) Capacity spectrum method of ATC-40

The second method is the 'capacity spectrum' method of ATC-40. This appears much more complex, but once mastered can provide a good insight into the processes and assumptions involved. The method uses the pushover curve to define an equivalent viscous linear system, with a secant stiffness corresponding to the target maximum displacement and a level of viscous damping related to the hysteretic damping in the real structure. The steps involved are (in outline)

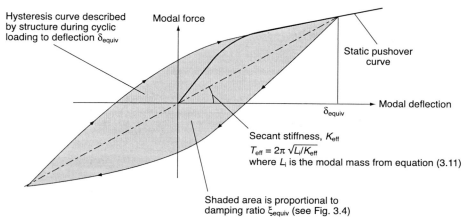

Fig. 3.24 Determining the equivalent SDOF period and damping from a pushover curve

as follows. (Skip to the next section if you wish to avoid the arguments involved.)

(1) Transform the forces and deflections in the pushover curve to modal quantities, in order to reduce the real MDOF system structure to an equivalent SDOF structure corresponding to the predominant mode of vibration. Referring to equations (3.11) and (3.12), it can be shown that this involves division of the forces and deflections by the modal factor

$$L_i/M_i \qquad (3.15)$$

Here, the subscript i refers to the predominant mode of deformation in the earthquake, which will usually be the first mode. A typical value of the modal factor for regular framed buildings is 1.25. Of course, the mode shape will change once the structure yields, and so the modal factors will also change slightly. However, this theoretical change is small compared to the other approximations involved in the method.

(2) Calculate the modal mass, L_i (equation (3.11)). For regular framed buildings, this is around 60% of the total mass for the first mode; once again it will change slightly when the structure yields.

(3) Make an estimate of the top deflection under the design ground motions. This can be based on the deflection the structure would have experienced, had it remained elastic, which (as noted above for the target displacement method) is often a good approximation. Divide this by the modal factor (L_i/M_i) to get the equivalent modal deflection δ_{equiv}.

(4) From the static pushover curve, find the slope of the secant stiffness corresponding to this equivalent modal deflection (Fig. 3.24).

(5) Calculate the period T_{equiv} and damping ξ_{equiv} of an equivalent linear SDOF system corresponding to this equivalent modal deflection (Fig. 3.25). Note that, with increasing deflection, both T_{equiv} and ξ_{equiv} increase.

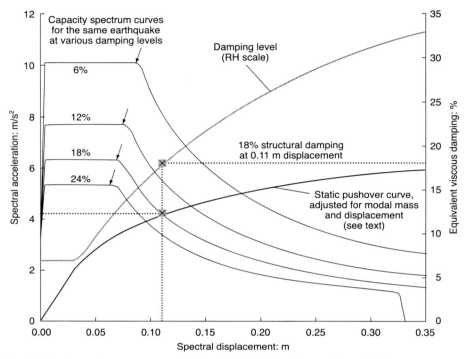

Fig. 3.25 Capacity spectrum graphical method for determining displacements

(6) From the response spectrum of the design ground motions, calculate the maximum displacement δ_{max} corresponding to T_{equiv} and ξ_{equiv}.
(7) If δ_{equiv} (the initial guess) differs from δ_{max} (the value found from the response spectrum), repeat step 2 with a modified value of δ_{equiv}, and iterate until satisfactory convergence is achieved.
(8) The top displacement of the real structure is then given as

$$\delta_{max}(L_i/M_i)$$

This analysis can be achieved more directly by plotting the design spectrum in the form shown in Fig. 3.8 – that is, as spectral acceleration against spectral displacement for various damping levels. The static pushover curve can also be plotted on this curve, provided the necessary transformations are made. First, it must be converted to modal quantities by dividing forces and deflections by the modal factor (L_i/M_i), as discussed before. The modal force must then be converted to a modal acceleration by dividing by the modal mass L_i.

The advantage of this method is that any point on the static pushover curve represents a particular value of structural period. A particular point on one of the capacity spectrum curves also represents a structural period (Fig. 3.8). Therefore, the point at which the pushover curve intersects one of the capacity spectrum curves represents a common structural period, displacement and acceleration demand. What is not necessarily in common, however, is the structural damping. By drawing a series of capacity spectrum curves at different damping levels, the damping level can be found where the intersection point implies a structural damping level achieved by the structure at its peak displacement. In the example of Fig. 3.25, the peak displacement is found by the intersection of the 18% damped spectrum with the pushover curve; the spectral acceleration is $4.2 \, \text{m/s}^2$ and the spectral displacement is $0.11 \, \text{m}$. By checking the damping level corresponding to this displacement from the appropriate curve, it can be seen that the structural damping does in fact equal 18%.

Having found the peak or spectral displacement of the equivalent SDOF structure, it must be reconverted to a top displacement of the real structure by multiplying by the modal factor, L_i/M_i (equation (3.15)).

(e) Interpretation of results

Both target displacement and capacity spectrum methods allow an estimate of the maximum seismic deflection to be made from a conventional response spectrum. This deflection can then be substituted back into the original static pushover analysis, and the corresponding degree of yielding in the structure can be established. For example, the rotation of plastic hinges in the beams can be found. These quantified measures of local yielding correspond to the degree of damage that the structure would experience, given the calculated maximum deflection. For example, the rotation of plastic hinges at the ends of beams provides a measure of local yielding. FEMA 356 provides guidance – see section 2.4. A typical table for reinforced concrete members forming part of a seismic resisting frame is shown in Table 3.4. Lower acceptance criteria would apply to members governed by shear

Table 3.4 *Typical performance criteria for concrete frame members resisting seismic loads*

	Acceptance criteria – plastic rotation angle in radians		
	Performance level		
	Immediate Occupancy (IO)	Life Safety (LS)	Collapse Prevention (CP)
Beams:			
Low shear, well confined	0.5–1%	1–2%	2–2.5%
High shear, well confined	0.5%	0.5–1%	1.5–2%
Low shear, poorly confined	0.5%	1%	1.0–2%
High shear, poorly confined	0.15%	0.5%	0.5–1%
Columns:			
Low axial load, well confined	0.5%	1.2–1.5%	1.6–2.0%
High axial load, well confined	0.3%	1.0–1.2%	1.2–1.5%
Low axial load, poorly confined	0.5%	0.4–0.5%	0.5–0.6%
High axial load, poorly confined	0.2%	0.2%	0.2–0.3%

(1) Establish performance objective, e.g. Life Safety (LS) performance under 475-year return event.

(2) Perform a non-linear static (pushover) analysis and establish a pushover curve.

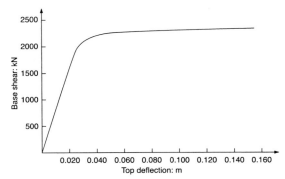

(3) Using the target displacement method (see subsection 3.4.3(c)) or capacity spectrum method (see subsection 3.4.3(d)), establish the top displacement of the structure under the design earthquake (e.g. 475-year return event).

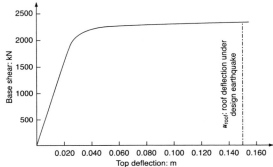

(4) Return to the non-linear static (pushover) model, and impose lateral forces so that the top deflection reaches the value under the design earthquake, calculated in step 3.

(5) Find the corresponding plastic deformations at the points which have yielded.

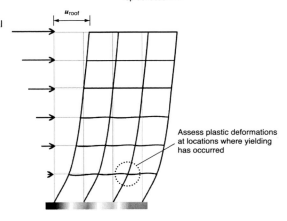

Assess plastic deformations at locations where yielding has occurred

(6) Check whether the plastic deformations exceed the limits for the chosen performance objectives. For example, the plastic rotations in the beams and columns can be compared with the limits in tables in FEMA 356.

| | | Maximum plastic rotation angle (radians) | | | | |
| | | Primary elements | | | Secondary elements | |
		IO	LS	CP	LS	CP
Beams	Ductile	1%	2%	2.5%	2%	5%
	Non-ductile	0.15%	0.2%	0.3%	0.5%	1%
Columns	Ductile, low axial load	0.5%	1.5%	2%	2%	3%
	Non-ductile, high axial load	0.2%	0.2%	0.2%	0.5%	0.8%

Typical plastic rotation limits from FEMA 356 for concrete beams for performance objectives IO (Immediate Occupancy), LS (Life Safety) and CP (Collapse Prevention).

Major advantages of method
- Information is provided on the *distribution* of plastic deformations (and hence damage) through the structure.
- Direct information is given on *local* plastic deformations, which can be related directly to performance state.

Some significant drawbacks
- The response depends on the distribution of horizontal forces assumed in the analysis, which is usually taken as unchanging. In reality, these driving inertia forces will change as the structure deforms plastically.
- The loading considered is in one direction only, instead of being cyclic as in an earthquake.
- The method becomes much less straightforward for buildings (e.g. those with torsional eccentricities) which require a 3-D model, rather than a 2-D model.
- Approximates an MDOF system to an SDOF system.

Fig. 3.26 Summary of displacement-based design procedure

failure rather than flexure, but higher criteria would apply to structural members not designed to resist seismic loads. Annexes A and B of Eurocode 8 Part 3 provide analytical expressions for acceptance criteria in steel and concrete elements, and Annex B of Eurocode 8 Part 2 gives an analytical method for concrete plastic hinges. These analytical methods may be useful for cases not covered by FEMA 356.

Figure 3.26 provides a brief summary of all the stages described above.

3.5   Analysis for capacity design

Ductile behaviour in a structure requires that yield capacity is reached first in ductile response modes (such as bending of well-detailed steel or concrete beams) rather than brittle modes (such as shear in poorly detailed concrete beams, buckling of slender steel struts or failure of welded connections). This design aim (known as capacity design) can be achieved by a suitable analysis to check that the requisite hierarchy of strength is present, implying that ductile modes are weaker than brittle modes. In essence, the brittle elements are designed to be strong enough to withstand the full *capacity* of the ductile, yielding elements – hence the term 'capacity design'.

An important concept in capacity design is that of 'overstrength'. The brittle members need to be strong enough to withstand the forces induced by yielding of the ductile members, allowing a suitable margin to give a high level of confidence that the brittle elements will not reach their failure loads. The overstrength of the yielding regions must allow for various possibilities, including strain hardening in steel, the possibility that actual strength on site is greater than specified strength and (sometimes) uncertainties in analysis. Moreover, the required strength in the brittle members must be based on the actual strength provided in the ductile elements; this almost always exceeds the minimum code requirement, for example the rounding up of member dimensions or bar diameters for practical reasons.

A straightforward example of capacity design is to check that the shear strength of a concrete beam in a frame under sway loading exceeds the force corresponding to the development of plastic hinges (Fig. 3.27).

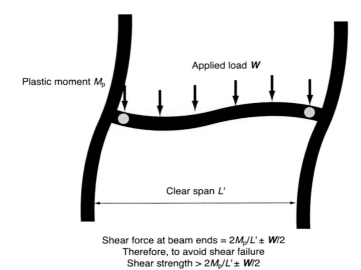

Shear force at beam ends = $2M_p/L' \pm W/2$
Therefore, to avoid shear failure
Shear strength $> 2M_p/L' \pm W/2$

Fig. 3.27 Capacity design for shear in a reinforced concrete beam

The design shear strength in this example follows in a statically determinate manner from the flexural strength at the plastic hinge points. Note however that in this example the hinges are assumed to form at the ends of the beam, which may not be the case for relatively high levels of applied gravity load.

Another example relates to the columns of unbraced sway frames. Here the aim is to ensure that yielding occurs first in the beams and not the columns, in order to achieve a 'strong column/weak beam' structure (Fig. 3.28(a)) and to avoid the soft or weak storeys.

The capacity design procedure is to ensure that the flexural strength of the columns framing into a joint exceed the sum of the plastic yield moments at the ends of the beams (Fig. 3.28(b)). There is more uncertainty here than for the previous case, because the distribution of moment between columns above and below any joint is not statically determinate; the ratio of $M_{\text{col, upper}}$ to $M_{\text{col, lower}}$ in Fig. 3.28(b) depends upon the points of contraflexure in the columns. Different codes treat the problem in different ways. The simplest approach is to require that the sum of the flexural strengths of the columns at each joint exceeds the sum of the beam flexural strengths by a suitable margin and this is essentially the IBC requirement, and is also a general requirement of Eurocode 8. The New Zealand concrete code NZS 3101 and its commentary provide the most detailed and complex procedure for concrete sway frames. This provides the greatest assurance that plastic hinges will not develop in columns during even the severest earthquake. The simpler procedures should prevent the formation of column hinges simultaneously at the top and bottom of a storey, and hence prevent a weak storey collapse, but may not prevent hinge formation at one end of a column.

The great advantage of capacity design, and the reason why it now finds a place in all major codes, is that it is a simple procedure which results in a ductile structure, more or less independently of detailed dynamic analysis procedures

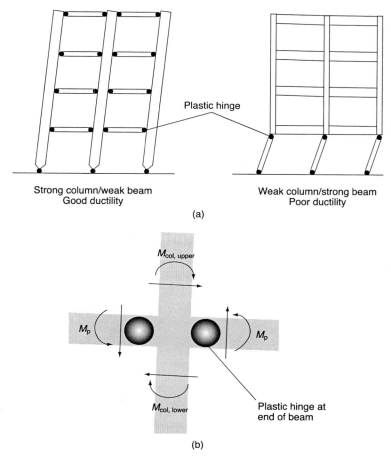

Fig. 3.28 Capacity design for unbraced frames: (a) overall view; and (b) forces at a beam–column joint (axial forces not shown)

or assumptions about the nature of the earthquake loading a structure may experience.

3.6   Analysis of building structures

3.6.1   Objectives

The objectives of the seismic analysis of a building structure are likely to include the following.

(1) To establish member strength requirements to prevent undue damage in frequent (lower-intensity) earthquakes.

(2) To establish ductility demands in members designed to yield in rare (extreme) earthquakes.

(3) To establish strength requirements in brittle members required to remain elastic in rare earthquakes.
(4) To calculate displacements, for the purpose of ensuring that non-structural elements such as cladding are suitably protected, preventing impact between adjacent structures and checking that P–delta effects are not significant.
(5) To establish the nature of dynamic design input to equipment mounted on the structure, for example machinery, storage tanks.

3.6.2   Methods of analysis

Suitable methods of analysis are provided in codes of practice; in general, the more complex and tall the building, the more stringent the analysis that is required. Regular buildings up to around 15 storeys in height can usually be designed using equivalent static analysis; tall buildings or those with significant irregularities in elevation (sudden changes in mass or stiffness with height) or plan (separation between the centres of stiffness and mass at any level) require modal response spectrum analysis. Non-linear static or dynamic analysis is becoming more common in design practice, and has for many years been mandatory in Japan for buildings taller than 60 m (Fitzpatrick 1992).

3.6.3   Analytical models

One-dimensional (1-D) (stick cantilever) computer models of buildings may have some attractions, because they are very quick and simple to run and may be suitable for initial studies. However, deriving appropriate shear and bending stiffnesses for the model is not straightforward, and with current levels of computing power, 2-D models are usually the starting-point. Generally, beam and column elements will be used, although shear walls may be modelled as plate elements. 2-D models of course have the limitation that they can only model response in the 2-D plane of analysis, and so effects such as biaxial bending in columns and torsional response cannot be captured. The added complexity of a 3-D model is needed in these cases.

For linear analysis, the 2-D and 3-D models will be similar to those used for static loads such as gravity, with one important exception. In addition to information about structural stiffness, the model must also have information on mass distribution. Without knowledge of the mass, it is impossible to calculate either the natural periods and mode shapes of the structure (and hence its dynamic properties), or the inertia forces arising from the earthquake. These must be added to the nodes as lumped mass elements with appropriate inertia properties. Many programs will automatically generate structural masses, based on information on cross-sectional area and density supplied as input data, but of course mass arising from elements such as cladding that are not modelled structurally must be calculated separately and added in.

For non-linear analysis, additional information must be added on the yield properties of elements. Often, yielding is assumed to be confined to predetermined plastic hinge regions. Advice on post-yield characteristics is given in FEMA 356 (FEMA 2000).

3.7 References

ASCE (1998). ASCE 4-98. *American Society of Civil Engineers Standard: Seismic analysis for safety-related nuclear structures.* ASCE, Reston, VA.

ATC (1996). ATC-40. *Seismic evaluation and retrofit of concrete buildings.* Applied Technology Council, Redwood City, CA.

Bruneau M. & Martin S. (1990). Inelastic torsional response of initially symmetric systems with load resisting elements having dissimilar strengths. *Proceedings of the 9th European Conference on Earthquake Engineering, Moscow.*

CEN (2004). EN1998-1: 2004. *Design of structures for earthquake resistance.* Part 1: General rules, seismic actions and rules for buildings. European Committee for Standardisation, Brussels.

Chandler A. (1990). Parametric earthquake response of torsionally coupled buildings and comparison with building codes. *Proceedings of the 9th European Conference on Earthquake Engineering, Moscow.*

Clough R. W. & Penzien J. (1993). *Dynamics of Structures.* McGraw Hill, New York.

Der Kiureghian A., Keshishian P. & Hakobian A. (1997). *Multiple support response spectrum analysis of bridges including the site-response effect and the MSRS code.* Report No. UBC/EERC-97/02, College of Engineering to the California Department of Transportation, University of California at Berkeley.

Ellis B. R. (1980). An assessment of the accuracy of predicting the fundamental natural frequencies of buildings and the implications concerning the dynamic analysis of structures. *Proceedings of the Institution of Civil Engineers*, Part 2, Vol. 69, Sept., pp. 763–776.

Fajfar P. (2000). A non-linear analysis method for performance-based seismic design. *Earthquake Spectra*, Vol. 16, Issue 3, August, pp. 573–592.

FEMA (2000). FEMA 356. *Prestandard and commentary for the seismic rehabilitation of buildings.* Prepared by the American Society of Civil Engineers for Federal Emergency Management Agency (FEMA), Washington DC.

FEMA (2005). ATC-55. *Improvement of inelastic seismic analysis procedures.* Prepared by the Applied Technology Council for the Federal Emergency Management Agency, FEMA 440, Washington DC.

Fenwick R. C. & Davidson B. J. (1987). Moment redistribution in seismic resistant concrete frames. *Proceedings of a Pacific Conference on Earthquake Engineering*, New Zealand, Vol. 1, pp. 95–106.

Fenwick R. C., Davidson B. J. & Chung B. T. (1992). *P*–delta actions in seismic resistant structures. *Proceedings of Bulletin of the New Zealand Society for Earthquake Engineering*, Vol. 25, pp. 56–69.

Fitzpatrick A. (1992). Structural design of Century Tower, Tokyo. *The Structural Engineer*, Vol. 70/18, September, pp. 313–317.

Gupta A. K. (1990). *Response Spectrum Method in Seismic Analysis and Design of Structures.* Blackwell Scientific, Cambridge, MA.

ICC (2003). IBC: 2003. *International Building Code.* International Code Council, Falls Church, VA.

Irvine H. M. (1986) *Structural Dynamics for the Practising Engineer.* Allen & Unwin, London.

Standards New Zealand (1995). NZS 3101: Parts 1 and 2: 1995. *The design of concrete structures.* Standards New Zealand, Wellington, NZ. (Note: A revised edition is expected to be published in 2006.)

4 Analysis of soils and soil–structure interaction

'Seismic loading is unique in that the medium (i.e. the soil) which imposes the loading on a structure also provides it with support.'

This chapter covers the following topics.

- Soil properties for seismic design
- Liquefaction: prediction and countermeasures
- Site amplification effects
- Topographical effects
- Slope stability
- Fault breaks
- Soil–structure interaction analysis

4.1 Introduction

The designer of earthquake-resistant structures needs some understanding of how soils respond during an earthquake; not only is this important for the foundation design itself, but the nature of soil overlaying bedrock may have a crucial modifying influence on the overall seismic response of the site. This chapter gives a fairly brief overview of soil properties under seismic excitation, and also reviews site response and soil–structure interaction effects. For a more detailed discussion of these issues, the reader is referred to Pappin (1991).

4.2 Soil properties for seismic design

4.2.1 Introduction

The response of soils to earthquake excitation is highly complex and depends on a large range of factors, many of which cannot be established with any certainty. The discussion that follows is intended to highlight the important features that apply to most standard cases; often, specialist geotechnical expertise will be needed to resolve design issues encountered in practice.

4.2.2 Soil properties for a dynamic analysis

In common with any structural system, dynamic response of soil systems depends on inertia, stiffness and damping. These three properties are now discussed in turn.

(a) Inertia

This can easily be determined from the soil's bulk density, which for most clays and sands is in the range 1700–2100 kg/m^3. There are exceptions, however; for example, Mexico City clay has a bulk density of only 1250 kg/m^3.

(b) Stiffness and material damping

Generally, the shear behaviour of soils will be of most concern; the behaviour in compression, characterised by the bulk stiffness, is less important. This is because the bulk stiffness of saturated materials is very high, being approximately equal to that of water divided by the soil porosity. For compression effects (for example, the transmission of P or seismic compression waves, important for vertical motions), the soil therefore acts in an essentially rigid manner with little modification due to dynamic effects. Soils with significant proportions of air may have much lower bulk stiffness, which may, therefore, need consideration. Further discussion here is confined to shear behaviour, which dominates response to horizontal seismic motion.

Figure 4.1 shows a typical cyclic response of a soil sample under variable-amplitude shear excitation. There are three important features to note when comparing the small with large shear strain response. First, the stiffness, determined from the slope of the stress–strain curve, decreases with shear strain. Second, the area contained within the hysteresis loop formed by the stress–strain curve increases with shear strain. As explained in Chapter 3, this area is directly related to the level of hysteretic damping. Therefore, soil damping increases with strain level, as more energy is dissipated hysteretically. It is important to note

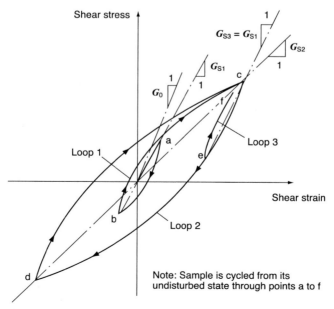

Fig. 4.1 *Idealised stress–strain behaviour of a soil sample in one-dimensional shear*

that the dissipated energy is generally much more dependent on amplitude than rate of loading. This is in contrast with viscous damping, where the damping resistance depends upon speed, and so for example reduces to zero for very slow rates of cycling. No such reduction to zero occurs in soils. Soil damping is thus essentially hysteretic in nature, which has important consequences for analytical modelling (see subsection 4.5.1).

A final feature to notice is that after a large shear strain excursion, the hysteresis loop reverts to its original shape for a small cyclic excitation; that is, loop 3 in Fig. 4.1 is similar in shape to loop 1, despite the intervening loop 2. Therefore, both stiffness and damping under cyclic loading are functions primarily of shear strain amplitude, not absolute shear strain.

4.2.3   Stiffness of sands and clays

Figure 4.2 shows typical relationships between shear strain amplitude and shear stiffness. Note the very large reduction in stiffness for shear strains exceeding 0.01%. The values for clays are for overconsolidation ratios (OCRs) of 1–15. It can be seen that the stiffness of clays becomes similar to that of sands as the plasticity index (PI) approaches zero.

In Fig. 4.2, the stiffness is expressed as a ratio of secant shear stiffness at the shear strain of interest, G_s, to the small strain stiffness, G_0. G_0 can be measured directly on site from measurements of shear wave velocity (see Pappin 1991) or from more conventional measurements, using empirical relationships. For sands, these relate G_0 to the blow count N for 300 mm penetration in the Standard Penetration Test (SPT); typical correlations between G_0 (in MPa) and blow count used in Japanese practice (Imai and Tonouchi 1982) are $G_0 = 7N$ and $G_0 = 14.4N^{0.68}$, but there is considerable scatter in the data. For clays, G_0 can be determined as a ratio of the undrained shear strength, c_u, as shown in Table 4.1.

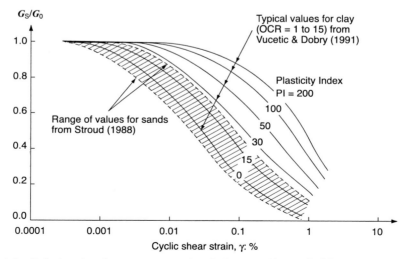

Fig. 4.2   Relationship between normalised shear stiffness G_s/G_0 and cyclic shear strain

Table 4.1 $\mathbf{G}_0/c_u$ *values (from Weiler 1988)*

Plasticity Index, PI: %	Overconsolidation ratio (OCR)		
	1	2	3
	$\mathbf{G}_0/c_u$		
15–20	1100	900	600
20–25	700	600	500
26–45	450	380	300

4.2.4 Material damping of sands and clays

Figure 4.3 shows typical values of damping ratio; once again, the values for clay approach those for sand as the PI reduces. Note the marked increase in damping as shear strains rise above 0.001%, caused by the hysteretic energy dissipation discussed in subsection 4.2.2. Stokoe *et al.* (1986) advise that the lower bound of the damping values shown for sands on the figure may be generally appropriate.

4.2.5 Stiffness and damping properties of silts

Silts have properties equivalent to clays with a PI of about 15% (Khilnani *et al.* 1982).

4.2.6 Strength of granular soils

The cyclic loading imposed on soils during an earthquake may seriously affect soil strength. Granular materials, such as sands and gravels, rely for their strength on

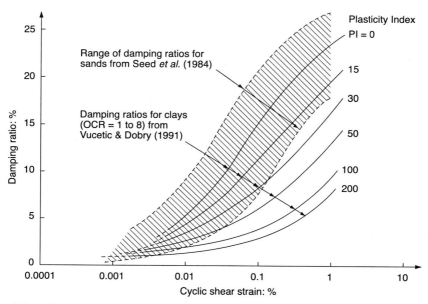

Fig. 4.3 Relationship between material damping ratio and cyclic shear strain

interparticulate friction. Although the angle of friction, ϕ', is not affected by cyclic loading, the effective stress between particles will be reduced in saturated soils if porewater pressures rise during an earthquake. The reduction in effective stress in turn reduces the shear strength. A rise in porewater pressure will occur if a loose granular material tries to densify under the action of earthquake shaking and the pressure has not had time to dissipate. In time, the porewater will find drainage paths, the pressure will release and the strength will be restored. This may however take a few minutes to occur, and dramatic failures can arise in the meantime (for example Fig. 1.23). This is the phenomenon of liquefaction, which is discussed more fully in section 4.3. The strength of granular soils is scarcely affected by the rate of loading.

4.2.7   Strength of cohesive soils

Clay particles are weakly bonded and are not subject to densification under cyclic loading. Therefore, they are unlikely to liquefy. The short-term undrained shear strength c_u however, is affected both by the rate of loading and by the number of cycles of loading. Rate effects may give rise to strength increases of up to 25% in soft clays under seismic loading conditions, compared with static strength, although the increase is less for firm clays and very stiff clays are insensitive to rate effects.

Strength reduction under cyclic loading is progressive with a number of cycles. It is highly dependent on the overconsolidation ratio (OCR). Clays with high OCR are much more sensitive to cyclic loading, and their strengths revert to normally consolidated values with increasing numbers of load cycles. The strength loss is permanent, unlike that due to porewater pressure increase in sands. A normally consolidated clay (OCR = 1) can sustain ten cycles of 90% of the undrained static shear strength c_u; this drops to ten cycles at about 75% c_u for a clay with OCR of 4 and to ten cycles at about 60% c_u for OCR of 10. Ten cycles of extreme loading is a very conservative estimate except in very large magnitude earthquakes.

4.3   Liquefaction

4.3.1   Assessing the liquefaction potential of soils

Liquefaction is a phenomenon which occurs in loose, saturated, granular soils under cyclic loading. Under such loading, porewater pressure between the soil particles builds up as the soil tries to densify, until the porewater pressure overcomes the forces between soil particles (Fig. 4.4) (i.e. the effective stress drops to zero). At this point, uncemented granular soils lose their shear strength, since this relies on interparticulate friction. Only certain types of soil are susceptible to liquefaction, and in order for it to occur, all the following features must be present

(a) a soil which tends to densify under cyclic shearing
(b) the presence of water between the soil particles
(c) a soil which derives at least some of its shear strength from friction between the soil particles
(d) restrictions on the drainage of water from the soil.

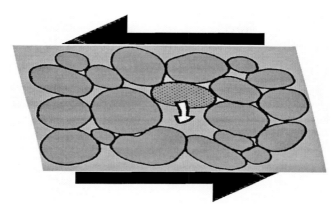

Fig. 4.4 Shearing of a loose, water-saturated, granular soil in the process of liquefying (modified version from EERI, 1994)

Condition (*a*) implies a loose soil; common examples are naturally deposited soils that are geologically young (Holocene deposits younger than 10 000 years) or man-made hydraulic fills. Densification and also cementation between particles (see condition (*c*)) tend to increase with age, and so older deposits are less susceptible to liquefaction. Conversely, land reclaimed by pumped dredged material is highly susceptible, unless suitable measures are undertaken. Table 4.2 provides a more detailed list of the susceptibility of soils.

Condition (*b*) necessitates that the soil is below the water table, although liquefaction is very unlikely where the water table depth is deeper than 15 m (Youd 1998).

Condition (*c*) means that granular soils are the most likely to liquefy, although silts still have some potential for liquefaction.

Condition (*d*) means that large-grained soils such as gravels are unlikely to liquefy, because any potential build-up of porewater pressure is usually dissipated rapidly by the free drainage available. As grain size decreases, the resistance to porewater drainage increases, but offsetting this is an increase in cementation between particles. The main risk of liquefaction therefore occurs in sands. However, silts may still liquefy, while coarse sands can liquefy if they are contained as lenses in larger areas of clay which inhibit dissipation of excess porewater pressures. Table 4.3 shows criteria developed by Seed and Idriss (1982) which are often used for a preliminary and usually conservative assessment of liquefaction, based on a soil's grading, moisture content and liquid limit.

4.3.2 Analytical methods of assessing liquefaction

Having established that a soil poses a potential liquefaction risk, the overall risk of it actually occurring must be related to the seismic hazard at the site; clearly the more intense the motions, the greater the risk. The most common method of calculation involves the following steps.

(*a*) The effective shear stress τ_e occurring in the soil during a design earthquake must first be calculated. τ_e corresponds to constant amplitude cyclic loading,

Table 4.2 Estimated susceptibility of sedimentary deposits to liquefaction (Youd 1998)

Type of deposit	Age of deposit			
	<500 years	Holocene	Pleistocene	Pre-Pleistocene
	Likelihood that cohesionless sediments, when saturated, would be susceptible to liquefaction			
(a) Continental deposits:				
River channel	Very high	High	Low	Very low
Floodplain	High	Moderate	Low	Very low
Alluvial fan and plain	Moderate	Low	Low	Very low
Marine terraces and plains	—	Low	Low	Very low
Delta and fan-delta	High	Moderate	Low	Very low
Lacustrine and playa	High	Moderate	Low	Very low
Colluvium	High	Moderate	Low	Very low
Talus	Low	Low	Very low	Very low
Dunes	High	Moderate	Low	Very low
Loess	High	High	High	Unknown
Glacial till	Low	Low	Very low	Very low
Volcanic tuff	Low	Low	Very low	Very low
Volcanic tephra	High	High	?	?
Residual soils	Low	Low	Very low	Very low
Sebka	High	Moderate	Low	Very low
(b) Coastal zone – delta and estuarine:				
Delta	Very high	High	Low	Very low
Estuarine	High	Moderate	Low	Very low
(c) Coastal zone – beach:				
High wave energy	Moderate	Low	Very low	Very low
Low wave energy	High	Moderate	Low	Very low
Lagoonal	High	Moderate	Low	Very low
Foreshore	High	Moderate	Low	Very low
(d) Artificial fill:				
Uncompacted fill	Very high	—	—	—
Compacted fill	Low	—	—	—

Table 4.3 Criteria for assessing liquefiability of fine-grained soils (based on Seed and Idriss 1982)

Criteria required for liquefaction of fine-grained soils (all three criteria must be met for soil to be liquefiable)
• Clay fraction (per cent finer than 0.005 mm) <15%
• Liquid limit (LL) <35%
• Moisture content (MC) >0.9 LL

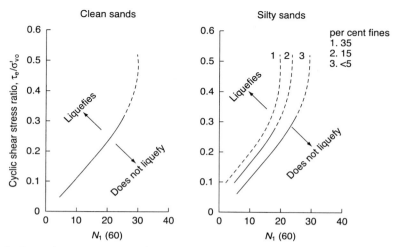

Fig. 4.5 Liquefaction potential for magnitude 7.5 earthquakes, based on SPT values (Eurocode 8, Part 5, CEN 2004)

and is generally taken as 65% of the peak value occurring during seismic loading, which allows for the fact that the peak occurs only once during the earthquake. A preliminary estimate of τ_e can be made from

$$\tau_e = 0.65 a_g \sigma_{vo}/g \qquad (4.1)$$

where a_g is the peak acceleration at ground level, after allowing for soil amplification effects (m/s^2); g is the acceleration due to gravity ($= 9.81$ m/s^2); and σ_{vo} is the vertical *total* stress at the level of interest (i.e. the total gravity overburden pressure).

Equation (4.1) assumes that the peak shear stress at the level of interest is $(a_g \sigma_{vo}/g)$. In fact, this is generally rather conservative, and a more rigorous analysis would use a simple one-dimensional shear beam model of the soil to estimate the peak cyclic shear stress on the soil at any depth, for example using SHAKE (1991), as discussed in the subsection on site amplification effects (4.4.1). The equivalent shear stress τ_e can then be taken as 65% of the peak value, since the peak occurs only once, as discussed above.

(b) τ_e is divided by the vertical *effective* stress σ'_{vo} at the level of interest (i.e. overburden stress less porewater pressure without allowance for liquefaction effects), to calculate the 'cyclic shear stress ratio', τ_e/σ'_{vo}.

(c) The liquefaction potential is then assessed as a function of the cyclic shear stress ratio, the type of soil and a soil property such as SPT (standard penetration test) value. Figure 4.5 shows the charts provided by Eurocode 8 Part 5 (CEN 2004). These are based on the corrected value of SPT blow-count in the soil N_{60}, which is calculated as explained in (d) to (f) below. It should be remembered that SPT is a relatively crude test, which depends on many things, including the test equipment and its operators, and the way the test borehole is drilled and backfilled.

Table 4.4 *Correction factors on critical value of shear stress ratio from Eurocode 8 Part 5 (CEN 2004) and Idriss (1999)*

	Correction factor for cyclic shear stress ratio	
Surface wave magnitude: M_s	EC8 Part 5	Idriss (1999)
5.0	Liquefaction unlikely	Liquefaction unlikely
5.5	2.86	1.69
6.0	2.20	1.48
6.5	1.69	1.30
7.0	1.30	1.14
7.5	1.00	1.00
8.0	0.67	0.88

(d) The SPT blowcount per 300 mm N_{SPT} is corrected to a standard value of effective vertical stress of 100 kPa by multiplying N_{SPT} by $(100/\sigma'_{vo})^{1/2}$, where σ'_{vo} is the effective vertical stress in kPa in the soil at the level of interest. EC8 advises that the correction factor should lie between the values 0.5 and 2.

(e) N_{SPT} is further corrected for energy ratio, by multiplying by $(ER/60)$ where ER is the percentage of the potential energy from the hammer drop which gets delivered to driving the SPT probe (the rest being lost in friction, noise, heat, rod vibration and so on). ASTM (1986) gives a method for quantifying ER, and further discussion is provided by Abou-Matar and Goble (1997).

(f) $N_1(60)$ in Fig. 4.5 is therefore given by the following equation

$$N_1(60) = N_{SPT}(100/\sigma'_{vo})^{1/2}(ER/60) \tag{4.2}$$

(g) Figure 4.5 relates to earthquakes of magnitude 7.5. The boundary value of cyclic shear stress ratio τ_e/σ'_{vo} at which liquefaction can be expected is calculated for other earthquake magnitudes by multiplying the Fig. 4.5 values by the correction factors in Table 4.4. Larger magnitude earthquakes tend to give rise to more cycles of loading, irrespective of the peak shear values arising, and Table 4.4 allows for this. Idriss (1999) proposes different values for these corrections factors, which suggest a lower dependence on earthquake magnitude.

(h) Eurocode 8 Part 5 suggests that the critical cyclic stress ratio from Fig. 4.5, at which the onset of liquefaction is expected, should be at least 25% greater than that estimated for the design earthquake (e.g. the 475-year return period event for most building structures).

These empirical correlations between SPT values and liquefaction potential suffer from the drawbacks of all empirical relationships. In particular, the SPT is a somewhat crude test, and measured SPT values depend on the details of the testing method, including the diameter and means of drilling the test boreholes. Therefore, the reliability of empirical predictions of liquefaction depends on the testing methods employed being similar to those used to derive the data shown in Fig. 4.5.

More sophisticated methods of assessing liquefaction risk have also been developed, whereby constitutive models of soil including porewater pressure generators are used in dynamic finite-element analysis. These models are still under development, and should always be supplemented by the more empirical measures described above.

4.3.3   Consequences of liquefaction

Having established that the soils around a structure may liquefy, the consequences must be evaluated.

The minimum consequence is that the densification associated with liquefaction gives rise to small local settlements, which may cause structural distress.

A much more serious consequence occurs where the reduction in shear strength caused by the liquefaction leads to a bearing failure (see for example Figs 1.1 and 1.23). Retaining walls are particularly at risk because they suffer not only from loss of bearing support but also from greatly increased lateral pressures, if the retained soil liquefies.

Lateral spreading can also occur, in which large surface blocks of soil move as a result of the liquefaction of underlying soil strata. The movements are usually towards a free surface such as a river bank, and are accompanied by breaking up of the displaced surface soil. Lateral spreading usually takes place on shallow slopes less than 3°. A dramatic example, which destroyed 70 houses, occurred during the Anchorage Alaska earthquake of 1964, when an area 2 km long by 300 m wide slid by up to 30 m (Fig. 4.6).

Fig. 4.6 Liquefaction-induced lateral spreading, Alaska 1964, showing destruction of a road and housing

The most catastrophic failure is a flow failure of soils on steep slopes (usually greater than 3°), which can give rise to displacements of large masses of soil over distances of tens of metres. The flows may be comprised either of completely liquefied soil, or of blocks of intact material riding on liquefied material (EERI 1994). Movements can reach tens of kilometres, and velocities can exceed 10 km/h.

Design measures in the presence of liquefiable soils are discussed in section 7.8.

4.4   Site-specific seismic hazards

The next subsections consider how the seismic hazard at a site may be affected by the local geology and how knowledge of the soil properties discussed in the previous sections can allow these hazards to be estimated.

4.4.1   Site amplification effects

The tendency of soft soils overlaying bedrock to amplify earthquake motions has already been discussed in Chapter 2, section 2.6. In many cases, adequate allowance for these effects can be made by simple amplification factors provided in codes of practice. It should be noted that amplification tends to reduce with increased intensity of ground motions because of the increase in soil damping and reduction in soil stiffness with shear strain amplitude (Figs 4.1–4.3). IBC: 2003 (ICC 2003) allows for this but Eurocode 8 (CEN 2004) does not, and this may be unconservative for soil sites where the peak ground acceleration is less than around 15%.

In cases where very soft materials are present, more sophisticated allowance should be made. Thus, at sites where soft clay layers are present which are deeper than 10 m and have a plasticity index PI > 40, Eurocode 8 requires a site-specific calculation of the modification they cause in surface motions. For horizontal motions, it is usually sufficient to make this modification on the basis of simple one-dimensional shear beam models of the soil, using the soil properties discussed in section 4.2. A range of bedrock motions appropriate to the site and to the depth of soil overlaying bedrock should be input to the base of the shear beam soil model and the ratio of surface to bedrock motion should be calculated at a range of frequencies. These frequency-dependent amplification factors can then be used to modify design spectra appropriate for rock sites. A number of standard computer programs exist to perform this calculation; SHAKE (1991) is a well-known example. The techniques are fully discussed by Pappin (1991).

One-dimensional shear beam models may not be adequate to describe site effects in alluvial basins where there is increasing evidence that more complex two- and three-dimensional effects are at work, particularly at the basin edges (Faccioli 2002). These effects are not currently addressed in codes of practice, and even complex finite-element modelling does not appear to yield reliable results (Adams and Jaramillo 2002).

The discussion so far has been on amplification of horizontal motions. Vertical motions are much less affected; they depend mainly on the bulk rather than the shear modulus of the soil, and since the former changes less than the latter (particularly in saturated soils) when the earthquake waves pass from rock into the overlying soil, little amplification occurs.

4.4.2 Topographical effects

Damage to structures is often observed to be greater on the tops of hills or ridges than at their base. An example was seen at a housing estate in Vina del Mar after the 1985 Chilean earthquake. Celebi (1987) measured ground motions during aftershocks of this event, both at the ridge-top positions, where damage had been greatest, and at the ridge base; he found that at certain frequencies the former motions were over ten times greater than the latter.

Eurocode 8 Part 5 provides for amplifications of up to 40% at the ridge of slopes greater than 15° forming part of a significant two-dimensional feature. Faccioli (2002) provides further information.

4.4.3 Slope stability

Slope failures connected with soil liquefaction were discussed in subsection 4.3.3. Even without liquefaction, the horizontal (and vertically upward) accelerations caused by an earthquake can dramatically reduce the factor of safety against movement of the slope. However, these reductions in factor of safety are instantaneous and only lead to large soil movements if the peak forces tending to displace the slope exceed the restraining strength of the soil by a factor of at least 2, i.e. where the instantaneous safety factor drops below 0.5. Relationships between instantaneous safety factor and slope displacement were originally developed by Newmark (1965) and form the basis for many current methods both of slope design and also for checking the seismic stability of retaining structures.

4.4.4 Fault breaks

Large earthquakes are almost always associated with rupture along fault lines. However, this rupture initiates at a depth of many kilometres and will rarely extend to the surface if the earthquake magnitude is 6 or less. Even for large earthquakes, a surface expression of the fault does not necessarily occur if large depths of soil overlay bedrock. The underlying fault movement (i.e. whether it consists of shear, tension or compression) also affects whether the fault reaches the surface.

For major active faults such as the San Andreas fault in California or the Northern Anatolian fault in Turkey which have a well-recorded history of movement, the design issues are clear: building structures should be sited away from them, and linear structures such as roads or pipelines should be designed to cope with possible fault movements. Generally, the width at risk should be taken as several hundred metres, allowing for the uncertainty in where the fault may appear at the surface in future earthquakes. However, the potential activity of other faults may be much harder to establish and not all potentially active faults have been mapped. For extended structures and systems such as pipelines or for very high-risk structures, further investigation may be needed (Mallard *et al.* 1991).

Structural damage from fault breaks arises not only from the consequences of straddling the fault (Fig. 1.2) but also the high pulses of ground motion ('seismic flings') that may arise in their vicinity (Bolt 1995). The seismic hazard maps for the USA provided in IBC (ICC 2003) allow for increases in ground motion of up to a factor of 2 in the vicinity of faults.

4.5 Soil–structure interaction

Most of the previous discussion has been based on response of soils in the 'free field' without man-made structures. The following subsections discuss briefly how to account for the interaction between a structure and its supporting soil.

4.5.1 Foundation flexibility

Structures founded on bedrock can be analysed assuming that their base is fixed. This assumption may be seriously in error, however, where the translational and rotational restraint offered to the structure by the soil is less than rigid. Usually, the effect of soil flexibility is to increase the fundamental period of the structure which often takes it away from resonance with the earthquake motions. Moreover, the cyclic movement of the soils in contact with the structure's foundations causes energy to be radiated away from the structure, tending to reduce its motion. This is known as radiation damping (Fig. 4.7). Generally, therefore, it is conservative to ignore these soil–structure interaction (SSI) effects, provided the site effects discussed in section 4.4 have been accounted for. The conservatism is not always present, however; Eurocode 8 Part 5 lists the following instances where SSI should be allowed for.

(a) Structures where **P**–δ effects (subsection 3.2.8) play a significant role.
(b) Structures with massive or deep-seated foundations, such as bridge piers, caissons and silos.
(c) Tall and slender structures such as towers and chimneys.
(d) Structures supported on very soft soils.
(e) The effect of the interaction between piles and the surrounding soils during earthquakes needs to be considered when the piles pass through interfaces between very soft soils and much stiffer soils.

An additional point is that even where none of these factors apply, structural deflections may well increase due to foundation flexibility; **P**–delta effects and

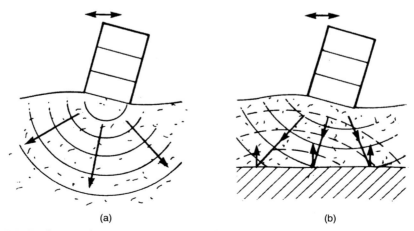

(a)   (b)

Fig. 4.7 Radiation damping: (a) waves radiating away from an oscillating building; and (b) reduction in radiation damping with thin soil strata

potential impact between structures may be adversely affected even if structural forces reduce.

A number of analytical techniques to investigate SSI are possible. The simplest method is to represent the soil flexibility by discrete springs connected to the foundation. For shallow foundations on deep uniform soils, the soil spring stiffness can be found from simple formulae; ASCE 4-98 (ASCE 1998) provides standard formulae for circular and rectangular bases. These require a knowledge of the shear stiffness of the soil, which, as shown in Fig. 4.2, depends on the shear strain amplitude. Where linear elastic analysis is performed, a series of iterative analyses is therefore required to find a suitable shear stiffness consistent with the computed shear strain. Similarly, Eurocode 8 Part 5 provides formulae for the effective stiffness of soil–pile systems.

The material damping associated with the soil spring is also strain-dependent (Fig. 4.2); a safe value for material damping of 5% is often taken. To this may be added the radiation damping, which may be significant. ASCE 4-98 provides values of equivalent viscous damping for uniform soils. These may be satisfactory where the soil depth is uniform over a depth much greater than the greatest foundation dimension. However, the presence of harder layers reflecting back radiated energy may significantly reduce radiation damping (Fig. 4.7(b)), and in this case special analysis is required.

In a response spectrum analysis, the damping levels due to material and radiation damping will apply only to the modes of vibration involving foundation movement, for which suitably reduced spectral accelerations can be assumed. Higher modes of vibration are unlikely to involve the foundation soils, so the damping level used should depend solely on the superstructure.

This type of analysis, assuming conventional linear springs, albeit modified in stiffness to allow for shear strain, may be satisfactory in many cases, but is theoretically not correct. A rigorous treatment of SSI effects using soil springs requires the use of springs whose stiffness and damping properties are frequency-dependent. Such an analysis can be relatively straightforward if frequency domain techniques are used. This type of analysis is discussed by Pappin (1991) and is not treated further here.

Finite-element modelling of soils is an alternative to the use of soil springs, and may be required to account for sloping or non-uniform soil strata, embedment of foundations and other complexities. The analysis is not straightforward, however, and there are special problems in treating boundaries of the portion of soil modelled in the analysis.

4.6 References

Abou-Matar H. & Goble G. (1997). SPT dynamic analysis and measurements. *Journal of Geotechnical and Geoenvironmental Engineering*, October, pp. 441–462. American Society of Civil Engineers, Reston, VA.

Adams B. M. & Jaramillo J. D. (2002). A two-dimensional study on the weak-motion seismic response of the Aburra Valley, Medellin, Colombia. *Bulletin of the New Zealand Society for Earthquake Engineering*, Vol. 35, Issue 1, March, pp. 17–41.

ASCE (1998). ASCE 4-98. *American Society of Civil Engineers Standard: Seismic analysis for safety-related nuclear structures*. ASCE, Reston, VA.

ASTM (1986). D4633-86. *Standard test method for stress wave energy measurement for dynamic penetrometer testing systems*. American Society for Testing and Materials, Philadelphia, USA.

Bolt B. A. (1995). *From earthquake acceleration to seismic displacement. The Fifth Mallet-Milne Lecture.* Wiley, Chichester, UK.

Celebi M. (1987). Topographical and geological amplifications determined from strong-motion and aftershock records of the 3rd March 1985 Chile earthquake. *Bulletin of the Seismological Society of America*, Vol. 77, Issue 4, pp. 1147–1167.

CEN (2004). EN1998-5: 2004. *Design of structures for earthquake resistance*. Part 5: Foundations, retaining structures and geotechnical aspects. European Committee for Standardisation, Brussels.

EERI (1994). *Earthquake basics: liquefaction – what it is and what to do about it.* Earthquake Engineering Research Institute, Oakland, CA.

Faccioli E. (2002). 'Complex' site effects in earthquake strong motion, including topography. *Keynote address, 12th European Conference on Earthquake Engineering*. Elsevier, Oxford.

ICC (2003). IBC: 2003. *International Building Code*. International Code Council, Falls Church, VA.

Idriss I. M. (1999). An update on the Seed–Idriss simplified procedure for evaluating liquefaction potential. In: *Proceedings, TRB [Transportation Research Board] Workshop on New Approaches to Liquefaction Analysis*. Available as Special Report Number MCEER-99-SP04 on: http://mceer.buffulo-edu/publications/workshop/99-SP04.

Imai T. & Tonouchi K. (1982). Correlations of N value with S-wave velocity and shear modulus. *Proceedings of the 2nd European Symposium on Penetration Testing*, May, Amsterdam, pp. 24–27.

Khilnani K. S., Byrne P. M. & Yeung K. K. (1982). Seismic stability of the Revelstoke earthfill dam. *Canadian Geotechnical Journal*, Vol. 19, pp. 63–75.

Mallard D. J., Higginbottom I. E., Muirwood R., Skipp B. O. & Wood R. M. (1991). Recent developments in the methodology of seismic hazard assessment. In: *Civil Engineering in the Nuclear Industry*. Institution of Civil Engineers, London, pp. 75–94.

Newmark N. M. (1965). *Effects of earthquakes on dams and embankments*. Selected papers. American Society of Civil Engineers, New York, 1976, pp. 631–652.

Pappin J. W. (1991). Design of foundation and soil structures for seismic loading. In: O'Reilly M. P. & Brown S. F. (eds) *Cyclic Loading of Soils*. Blackie, London, pp. 306–366.

Seed H. B. & Idriss I. M. (1982). *Ground Motions and Soil Liquefaction During Earthquakes*. Earthquake Engineering Research Institute, Oakland, CA.

Seed H. B., Wong R. T., Idriss I. M. & Tokimatsu K. (1984). *Moduli and Damping Factors for Dynamic Analysis of Cohesionless Soils*. College of Engineering, University of California at Berkeley, CA, Report No. UCB/EERC-84/14.

SHAKE (1991). A computer program for conducting equivalent linear seismic response analysis of horizontally layered soil deposits. Center for Geotechnical Modeling, University of California, Davis, CA.

Stokoe K. H., Kim J., Sykora D. W., Ladd R. S. & Dobry R. (1986). Field and laboratory investigations of three sands subjected to the 1979 Imperial Valley earthquake. *Proceedings of the 8th European Conference on Earthquake Engineering, Lisbon*, 8–12 September, Vol. 2, pp. 5.2/57–5.2/64.

Stroud M. (1988). The standard penetration test: its application and interpretation. *Proceedings of the Conference on Penetration Testing in the UK*, Institution of Civil Engineers, Birmingham, 6–8 July, pp. 29–49.

Vucetic M. & Dobry R. (1991). Effect of soil plasticity on cyclic response. *ASCE Journal of Geotechnical Engineering*, Vol. 117, Issue 1, pp. 89–109.

Weiler W. A. (1988). Small-strain shear modulus of clay. *Earthquake Engineering and Soil Dynamics II – Recent Advances in Ground Motion Evaluation. ASCE Geotechnical Special Publication No. 20*, pp. 331–345.

Youd T. L. (1998). *Screening guide for rapid assessment of liquefaction hazard at highway bridge sites*. National Center for Earthquake Engineering Research, Buffalo, NY, Technical Report MCEER-98-0005.

5   Conceptual design

'[Earthquake] safety can be expensive if you start off with
the wrong system, or architectural or engineering design.'

Henry Degenkolb. *Connections*.
Earthquake Engineering Research Institute,
Oakland, CA, 1994

This chapter covers the following topics.

- The anatomy of a building
- Planning considerations and overall form
- Framing systems
- Costs

5.1   Design objectives

Nothing within the power of a structural engineer can make a badly conceived
building into a good earthquake-resistant structure. Decisions made at the conceptual stage are difficult to modify so that it is essential that their full consequences
are understood in terms of performance and costs as early as possible.

5.2   Anatomy of a building

The functioning parts of a building affect the way in which it can accommodate its
structural skeleton. For this reason it is useful to consider the principal division of
functions and how they affect the structure.

The principal categories of building use can be considered in a vertical direction
as given in Table 5.1. Vertical divisions of function within the building may be a
source of problems, making it difficult to avoid irregularities in mass or stiffness.
For example, the ground floor of many commercial buildings is often taller and
more open than higher floors. However, the service cores and exterior cladding
provide an opportunity to incorporate shear walls or braced panels to overcome
resulting problems. One of the main objectives in early planning is to establish
the optimum locations for service cores and for stiff structural elements that will
be continuous to the foundation.

It is not unusual to find that structural and architectural requirements are in
conflict at the concept planning stage but it is essential that a satisfactory compromise is reached at this time.

Table 5.1 Categories of function within a building

Basement	Car parking, storage, mechanical and electrical plant
Street level	May be used quite differently from the rest of the building, commonly leading to a greater than typical storey height and a need for unobstructed floor space: for example, in hotels the street level may be used for reception, conference and restaurant areas in contrast with the regular pattern of rooms on the typical floors; in office buildings the street level may include shops, banks, restaurants, etc.
Typical floors	Repetitive standard levels
Roof structures	Mechanical and electrical plant, lift motor room, water tanks, etc.
Service cores	Stairs, lifts, toilets and pipe ducts, which are frequently grouped together
Usable floor	Clear spaces, usually modular
Exterior cladding	Provides opportunities for bracing, shear walls

5.3 Planning considerations

5.3.1 The influence of site conditions

It is essential to obtain data at an early stage on the soil conditions and ground-water level at the site, since these can have a major influence on seismic design. The principal aspects to determine are the period range over which the soils may amplify seismic motions, the liquefaction potential of the soil and the stability of slopes at the site. Initially at least, standard tests suffice, comprising in-situ tests (standard penetration test (SPT) or cone penetration test (CPT) values and groundwater level measurements) and laboratory tests (soil description and standard strength tests); additional specialist techniques such as shear wave velocity in-situ tests and cyclic triaxial or resonant column laboratory tests may be needed in special circumstances (e.g. soil profiles S_1 and S_2 in Table 5.2). Unless the soils at the site are well understood from previous investigations, borehole data to at least 30 m (or bedrock depth if less) are required.

For other than minor projects, the soil data need to be sufficient to classify the site into one of the standard profiles described in codes of practice. Table 5.2 shows the Eurocode 8 classification system, together with the period range for peak amplification of ground motions. Structures falling into this period range will be particularly highly stressed. As a rough initial guide, the fundamental period of a building is $N/10$. Consequently deep, soft soil deposits can be damaging to tall buildings, but also shallow, stiff deposits can prove troublesome for low-rise structures. If the building period is similar to that of the proposed structure, large amplification of seismic response will result and it may be worth considering ways of modifying the structural period to detune it from the earthquake motions. Increasing the stiffness (e.g. addition of bracing or shear walls) or reducing the mass (e.g. lightweight floors, lightweight concrete) both reduce the structural period, and of course the reverse is also true. However, period depends on the square root of mass divided by stiffnesses, so large changes in mass and stiffness

Table 5.2 Soil classification (from Eurocode 8, CEN 2004)

Classification	Description	Characteristic parameters in top 30 m		Period range T_B to T_C for peak ground motion amplification			
		Non-cohesive soils	Cohesive soils	Large earthquakes govern		Small earthquakes govern	
		N_{SPT} blowcount/300 mm	c_u (kPa) undrained shear strength	T_B: s	T_C: s	T_B: s	T_C: s
A	Rock or other rock-like geological formation, including at most 5 m of weaker material at the surface	—	—	0.15	0.4	0.05	0.25
B	Deposits of very dense sand, gravel, or very stiff clay, at least several tens of metres in thickness, characterised by a gradual increase of mechanical properties with depth	>50	>250	0.15	0.5	0.05	0.25
C	Deep deposits of dense or medium dense sand, gravel or stiff clay with thickness from several tens to many hundreds of metres	15–50	70–250	0.20	0.6	0.1	0.25
D	Deposits of loose-to-medium cohesionless soil (with or without some soft cohesive layers), or of predominantly soft-to-firm cohesive soil	<15	<70	0.20	0.8	0.1	0.3

	Description				
E	A soil profile consisting of a surface alluvium layer similar to type C or D and thickness varying between about 5 m and 20 m, underlain by stiffer material with a shear wave velocity >800 m/s	0.15	0.5	0.05	0.25
S$_1$	Deposits consisting – or containing a layer at least 10 m thick – of soft clays/silts with high plasticity index (PI >40) and high water content <100 (indicative)	10–20	Special investigations needed		
S$_2$	Deposits of liquefiable soils, of sensitive clays, or any other soil profile not included in types A–E or S$_1$	Special investigations needed			

are needed for a significant change in period. By contrast mounting the building on flexible bearings can dramatically increase the period.

Liquefaction or slope stability problems could lead to the conclusion that the site is unsuitable for development without expensive soil improvement measures or foundation solutions. An initial indication of the potential for soil liquefaction can be obtained from Table 4.2 in Chapter 4.

5.3.2   Structural layout

The experience of past earthquakes has confirmed the commonsense expectation that buildings which are well tied together and have well-defined, continuous load-paths to the foundation perform much better in earthquakes than structures lacking such features.

The degree of symmetry also has a significant influence on earthquake resistance. Earthquake damage is found to be five to ten times worse in buildings with significant irregularity, compared to those with essentially regular structures. The reason is that sudden changes in section cause stress concentrations and potential failure points. The most common example is the 'weak storey', often caused by architectural requirements for openness at ground-floor level. The result is that deflections are concentrated at this level during the earthquake, giving rise to very severe structural demands. Weak storeys have caused perhaps more collapses in earthquakes than any other feature. Another important example of irregularity occurs where the centres of stiffness and mass of a structure do not coincide, giving rise to damaging coupled lateral/torsional response. Compact plan shapes are also favoured, since flexible extensions from a structure are prone to vibrate separately from the rest of the structure.

The layout of the lateral load-resisting vertical elements should therefore aim for the greatest possible regularity, compactness and torsional resistance. Irregular plan shapes can be divided into compact shapes by providing separation joints; these must be sufficiently wide (up to 50 mm for each storey height above ground in flexible structures) to prevent damaging contact from occurring as the separated parts of the building sway in an earthquake.

Adequate separation is even more important between adjacent buildings, because the storey heights are unlikely to coincide, and a stiff floor diaphragm of one building may impact the other at the vulnerable mid-height position of columns.

The mass distribution within a building should also be considered. The characteristic swaying mode of a building during an earthquake implies that masses placed high in the building produce considerably more unfavourable effects than masses placed lower down. Massive roofs and heavy plant rooms at high level are therefore to be avoided where possible.

Finally, undue reliance on a few elements to provide lateral resistance should be avoided, since there is no backup if they fail. The combination of shear walls with moment frames is one way of ensuring such redundancy.

A classic work (Arnold & Reitherman 1982) on the need for symmetry in seismic conditions, written by two architects over 20 years ago, is still worth reading and sharing with architectural colleagues.

5.3.3 Ductile and brittle responses

All of these considerations contribute towards obtaining a good earthquake-resistant design, but do not necessarily ensure that there is adequate reserve to meet an extreme earthquake attack without collapse. The strategy commonly adopted is therefore to provide sufficient strength to minimise damage in an earthquake with a high probability of occurrence, but to accept that the structure may yield in a low-probability event with the accompanying risk of damage, while ensuring that the post-yield response is ductile rather than brittle. A ductile structure is one that can maintain its stability under repeated cyclical deflections considerably greater than its yield deflection. The ductile structure therefore resists the extreme earthquake not by brute force, but by allowing plastic deformations to absorb the kinetic energy induced by the ground shaking. The plastic yielding not only absorbs energy, but also softens the structure and increases its natural period, which will usually further reduce demand.

This strategy implies that considerable structural damage may occur in an extreme event, possibly to the extent that the structure is not repairable. Provided this has been assessed as a low-probability occurrence, and provided life safety is not impaired, this can be justified. Given the huge uncertainties both in predicting earthquake motions and calculating response, the provision of ductility is the best insurance policy against destruction of human lives.

Modern earthquake codes take advantage of ductile yielding to reduce the level of seismic design force, typically to a level two to eight times lower than the strength required for the structure to remain elastic. Lower ductility demands are implicit in Japanese practice, but reductions by a factor of 2 or more on elastic demands during a major earthquake are still permitted. This emphasises the point that provision of adequate strength is not in itself sufficient; measures to ensure ductility are also essential.

There are two principal means of ensuring ductility. First, the capacity design procedures, described in section 3.5, should be used to ensure that yielding takes place in ductile rather than brittle modes. Second, special detailing is needed to ensure that parts of the structure designed to yield can achieve large post-yield strains. An example is the provision of horizontal confinement steel in columns. The provisions of seismic codes of practice are much concerned with such details.

5.3.4 Provision of adequate stiffness

Deflections must be limited during earthquakes for a number of reasons, and hence provision of adequate stiffness is important. Relative horizontal deflections within the building (e.g. between one storey and the next, known as storey drift) must be limited. This is because non-structural elements such as cladding, partitions and pipework must be able to accept the deflections imposed on them during an earthquake without failure. Failure of external cladding, blockage of escape routes by fallen partitions and ruptured firewater pipework all have serious safety implications. Moreover, some of the columns in a building may only be designed to resist gravity loads, with the seismic loads taken by other elements, but if deflections are too great they will fail through 'P–delta' effects (subsection 3.2.8) however ductile they are.

Overall deflections must also be limited to prevent impact, both across separation joints within a building and (usually more seriously) between buildings.

For all these reasons, it is therefore essential to check that the stiffness of the structure conforms with code requirements; this criterion, rather than strength, often governs section sizes in tall buildings.

5.3.5   Interaction between structure and non-structure

The swaying of a building in an earthquake gives rise to inertial forces in a building's contents, just as much as in its structure, and failure to provide simple lateral restraint, for example to the racking in warehouses, has caused considerable damage in the past. Similar failures have occurred where mechanical and electrical plant, false ceilings, internal partitions, etc., have not been properly restrained, or where they are unable to accept the relative deflections imposed upon them.

Design of non-structural elements and their attachment is often dealt with at the detailing stage. However, interaction of the structure with stiff non-structural elements such as infill blockwork partitions or cladding elements can result in significant and often deleterious changes to structural response. At an early stage, it should be decided whether cladding, partitions, staircases and so on are to remain separate from the main structure or to be designed to work with it in resisting seismic loads. Liaison between the structural engineer and other members of the design team, such as architects and mechanical engineers, is essential to ensure safe seismic interaction between structure and non-structure.

5.4   Structural systems

5.4.1   Foundations

General guidance on the choice of foundation system is difficult, since the relative cost and efficiency of different types depend critically on the soil conditions and type of superstructure. Some factors that should be considered in connection with seismic resistance are as follows.

(a) Where the superstructure is designed to achieve a high level of ductility, the foundation must be able to develop the superstructure's yield capacity. It is no use having a perfectly detailed ductile superstructure supported by a foundation which suffers brittle failure before that ductility is achieved.

(b) Superstructure systems that involve large uplift forces (e.g. shear walls with a high height-to-width ratio) are only suitable if foundations can be built economically to resist these tension forces.

(c) Piles have loads imposed upon them due to lateral deflection of the upper layers of softer soil during earthquakes. Small driven piles of less than 0.5 m diameter are generally sufficiently flexible to accept this movement without suffering large bending stresses. Large-diameter piles, however, may experience significant lateral forces as they are relatively stiff compared with the soil.

(d) Raking piles are generally to be avoided because they add greatly to the lateral stiffness of the pile group. Their stiffness means that they will not be able to conform to the deformations of the soft soil strata, but will receive very large lateral loads, arising from the mass of the soft soils attempting to move past the stiffened pile group. Raking piles have been found to be prone to failure during earthquakes.

(e) Piling through potentially liquefiable layers needs careful consideration, since the piles would have to transmit the lateral forces from both the superstructure and adjacent non-liquefied soil through the liquefied strata. The piles would be effectively unsupported laterally in this region and so may be subject to large bending and shear stresses which would be difficult to resist.

(f) Raft foundation support via a basement may be an alternative solution when founding on potentially liquefiable layers, as discussed in section 7.8.

5.4.2 Choice of structural material

The most appropriate structural material to use is influenced by a host of different factors, including relative costs, locally available skills, environmental, durability, architectural considerations, and so on. Some of the seismic aspects are as follows.

Steel has high strength-to-mass ratio, a clear advantage over concrete because seismic forces are generated through inertia. It is also easy to make steel members ductile in both flexure and shear. However, providing adequate seismic resistance of connections can be difficult, and buckling modes of failure lack ductility.

Concrete has an unfavourably low strength-to-mass ratio, and it is easy to produce beams and columns which are brittle in shear, and columns which are brittle in compression. However, with proper design and detailing, ductility in flexure can be excellent, ductility in compression can be greatly improved by provision of adequate confinement steel and failure in shear can be avoided by 'capacity design' measures. Moreover, brittle buckling modes of failure are much less likely than in steel. Although poorly-built concrete frames have an appalling record of collapse in earthquakes, concrete shear wall buildings have an excellent record, even where design and construction standards are less than perfect.

Masonry, too, suffers from a high strength-to-mass ratio, and does not exhibit ductile failure modes but good-quality stone is very strong in compression. Where this compressive strength is harnessed to resist earthquake forces, particularly through the use of arches and domes, the performance is found to be good. Unlike the traditional engineering materials of steel and concrete, masonry structures must be designed elastically to have a large reserve against design earthquake forces, without reliance on ductility.

Timber (favourably) is strong and light, and its connections usually provide good levels of damping without suffering from the low-cycle fatigue problems that beset steel. However, timber can lose strength through fungal or insect attack, and it also burns easily.

5.4.3 Moment-resisting frames

(a) General characteristics of moment-resisting frames

Moment-resisting (i.e. unbraced) frames derive their lateral strength, not from diagonal bracing members, but from the rigidity of the beam–column connection.

They consist solely of horizontal beams and vertical columns, and are in common use for both steel and concrete construction.

The advantages of using moment-resisting frames to provide seismic resistance are as follows.

(1) Properly designed, they provide a potentially highly ductile system with a good degree of redundancy, which can allow freedom in the architectural planning of internal spaces and external cladding, without obstruction from bracing elements.

(2) Their flexibility and associated long period may serve to detune the structure from the forcing motions on stiff soil or rock sites.

The potential problems associated with moment-resisting frames are as follows.

(1) Poorly designed, reinforced concrete moment-resisting frames have been observed to fail catastrophically in earthquakes, mainly by formation of weak storeys and failures around beam–column junctions. Steel moment-resisting frames have performed better, but still proved vulnerable at welded connections.

(2) The beam–column joint region represents an area of high stress concentration which needs considerable skill to design successfully. In concrete, this often involves congested reinforcement, which needs good steel-fixing skills and good concreting to ensure proper compaction around the reinforcement. In steel, careful detailing of the connections and panel zone is needed. Where these skills do not exist, use of ductile moment-resisting frames is best avoided.

(3) The low stiffness of moment-resisting frames tends to cause high storey drifts (interstorey deflections), which may lead to a number of problems. These include unacceptable damage to cladding and other non-structural elements and to other serious structural problems. Moreover, the width of separation joints within the structure may need to be large to prevent buffeting during an earthquake, and this can lead to problems in detailing an acoustic, thermal and weathertight bridge to cross the joints. A more general problem with the flexibility of moment frames, particularly in tall buildings, is that design may be governed by deflection rather than strength, leading to an inefficient use of material.

Frames with overall height-to-base-width ratios of up to 4 are in common use. When used as the sole seismic-resistant system, higher ratios may result in uplift problems, particularly at corner columns, which tend to carry the lowest gravity load and attract the highest tensions due to lateral loads. Very slender structures are prone to deflection problems, both in excessive storey drifts and overall movement. The maximum practicable ratio depends, however, on the seismicity of the area and hence the magnitude of lateral forces that must be resisted. Moreover, wind loads as well as seismic loads need to be considered when choosing the overall slenderness.

The ratio of beam span to column height depends on a number of considerations. Internal frame geometries may be governed by the need for unrestricted internal spaces; optimisation of the structure supporting gravity loads may also result in a larger span than would be chosen for purely seismic resistance. External frames may be less restricted in this way; the optimum beam span is likely to be 1 to $1\frac{1}{2}$ times the storey height, although a wide range of ratios is found in practice.

The requirements of the governing code of practice may also influence the choice of structural system. Under the requirements of the Californian code IBC (ICC 2003), use of ductile moment-resisting frames to resist at least part of the seismic loads is mandatory in areas of high seismicity for tall buildings. There are no such restrictions in other codes such as Eurocode 8 or the New Zealand seismic codes.

(b) Grid frames and perimeter frames (Fig. 5.1)

Moment frames can be classified into two different types. The first, grid frames, comprise a uniform grid of frames in both directions, and are common in Japanese practice in low- to medium-rise construction. They are highly redundant (a favourable feature), and achieve a good spread of resistance to seismic forces both within the superstructure and to the foundations. They have very good torsional resistance and coupled lateral/torsional response is unlikely to be a problem, even with irregular plan shapes.

The major disadvantages are as follows. All the columns have to be designed for biaxial loads and all beams and columns have to be designed and detailed for ductility. External columns (especially corner columns) carry the lowest gravity loads, but are subject to the largest seismic axial forces and so there may be uplift problems. A grid frame may restrict to some extent the freedom of architectural planning of the internal space of a building.

Grid frames generally find their application in low- to medium-rise buildings, with any plan shape.

In the second type of system, perimeter frames, the seismic-resisting frames are restricted to the outside of the building. The interior space only needs structure

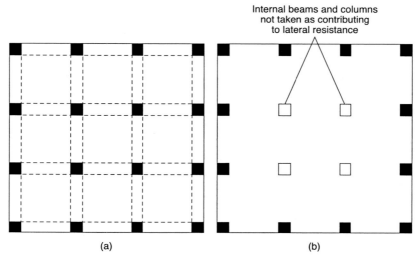

Fig. 5.1 Types of moment-resisting frame: (a) grid frame; and (b) perimeter frame

capable of supporting gravity loads; consequently, column spacing can be increased, allowing greater architectural freedom and, probably, economy. The good torsional stiffness of grid frames is retained, as is some of their redundancy.

The corner columns of perimeter frames in rectangular buildings suffer from the problems of biaxial loading and possible uplift referred to above for grid frames. Circular plan shapes are less affected by this problem.

Perimeter frames find their application in medium- to high-rise structures with compact plan shapes. High-rise frames are likely only to be economic in steel.

(c) Precast concrete frames

Precasting offers the general advantages of speed of erection, minimisation of costly formwork and falsework, and the improvement in quality control possible under shop fabrication conditions. The potential seismic problems (displayed so dramatically during the Armenian earthquake of 1988 – see Wyllie and Filson 1989) are the difficulties in ensuring ductility and continuity at the connections between precast units; the elements must be joined together so that they do not shake apart during a major earthquake. Extensive research and development in New Zealand has made the industry there confident that, properly designed and built (which the Armenian precast frame buildings certainly were not), precast frames can be safe in earthquake-prone regions, and they are commonly used in New Zealand for buildings of up to 20 storeys. They are found in Eastern Europe and increasingly in Japan, where they have been the subject of a major research effort. Their use is less common in the seismic areas of the USA or Western Europe.

(d) Blockwork infill in moment-resisting frames

Rigid blockwork infill of external moment-resisting frames offers a good solution for providing thermal and acoustic insulation and weatherproofing. The block-work infill causes a large increase in strength and stiffness, at the expense of a reduction in ductility, and there is evidence from recent earthquakes that such infill has protected poorly designed frames from collapse. However, if the infill is not uniform across the building, unsafe conditions such as the creation of a weak storey can result.

Unreinforced blockwork is not permitted in seismic areas of the USA. However, it is permitted in Eurocode 8, and design rules are presented, based on the long experience of its use in seismic parts of Southern Europe – another example of the influence of codes on an important aspect of conceptual design. The danger with its use is that panels may fail, creating a risk not only from falling masonry but also formation of a weak storey.

Provided a designer takes account of the interaction between infill and frame, allows for the reduction of ductility by provision of additional strength, and ensures that the masonry cannot fall out of the frame during an earthquake, rigid infill of frames with unreinforced masonry can be considered (at least if Eurocode 8 (CEN 2004) is the governing code). Otherwise, the blockwork must be separated from the frame so that it does not attract seismic forces, but the frame still provides restraint to prevent it from falling out during strong ground motion. This is a tricky detailing problem, but results in a more predictable and ductile system.

5.4.4 Concentrically braced frames (CBFs)

CBFs are conventionally designed braced frames in which the centre lines of the bracing members cross at the main joints in the structure, thus minimising residual moments in the frame (Fig. 5.2). The pros and cons of braced frames are essentially the opposite of moment frames; they provide strength and stiffness at low cost but ductility is likely to be limited and the bracing may restrict architectural planning.

Figure 5.2 distinguishes between various types of braced frame, the seismic resistance of which can be markedly different. Because of the cyclic nature of seismic loading, their behaviour under extreme lateral loads in alternating directions must be considered.

An X-braced frame (Fig. 5.2(a)) has bracing members in tension for both directions of loading, and if these are sized to yield before the columns or beams fail, ductility can be developed. However, after a brace has yielded in tension due to loading in one direction, it is liable to buckle rather than yield in compression on the reverse cycle. Plastic tensile strains therefore tend to accumulate in the braces, limiting the ductility that is achievable. Moreover, this accumulated tensile strain creates a slack in the system, because after one complete cycle of loading, the deflection needs to exceed the plastic excursion achieved in the previous load cycle before the tensile brace is loaded (Fig. 5.3). These effects become more marked (and unfavourable) with increasing slenderness of brace; IBC (ICC 2003) therefore distinguishes between 'special' and 'ordinary' CBFs, the former having limits on

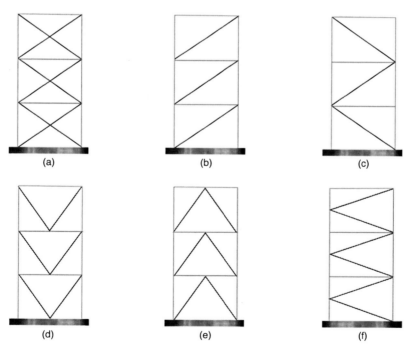

Fig. 5.2 Examples of bracing schemes for concentrically braced frames: (a) X-braced; (b) diagonally braced; (c) alternative diagonally braced; (d) V-braced; (e) inverted V-braced; and (f) K-braced

Fig. 5.3 Failure of X-braced steel frame, Kobe, Japan, 1995

slenderness ratio, but with reduced strength requirements reflecting the increased ductility available.

Single bays of diagonal braces (Fig. 5.2(b) and (c)) respond differently according to the direction of loading. Configuration (b) may be much weaker and flexible in the direction causing compression in the braces, while configuration (c) will be weaker and more flexible in the storeys with compression braces, leading to the possibility of soft-storey formation. This is clearly not satisfactory. With more than one diagonally braced bay, the performance can revert to that of X-bracing if a suitable arrangement of bracing direction is chosen. Eurocode 8 requires a balance of compression and tension braces at each level.

The V-braced arrangements of Fig. 5.2(d) and (e) suffer from the fact that the buckling capacity of the compression brace is likely to be significantly less than the tension yield capacity of the tension brace. Thus there is inevitably an out-of-balance load on the horizontal beam when the braces reach their capacity, which must be resisted in bending of the horizontal member. This restricts the amount of yielding that the braces can develop, and hence the overall ductility. Where the horizontal brace has a large bending strength which can resist the out-of-balance load, the hysteretic performance of V-braced systems is improved.

The same out-of-balance force applies to K-braces (Fig. 5.2(f)) when the braces reach their capacity, but this time it is a much more dangerous horizontal force applied to a column – dangerous because column failure can trigger a general collapse. For this reason, K-braces are not permitted in seismic regions.

5.4.5 Eccentrically braced frames (EBFs) and knee-braced frames
In EBFs, some of the bracing members are arranged so that their ends do not meet concentrically on a main member, but are separated to meet eccentrically (Fig. 5.4).

The eccentric link element between the ends of the braces is designed as a weak but ductile link which yields before any of the other frame members. It therefore provides a dependable source of ductility and, by using capacity design principles, it can prevent the shear in the structure from reaching the level at which buckling occurs in any of the members. The link element is relatively short and so the elastic response of the frame is similar to that of the equivalent CBF. The arrangement thus combines the advantageous stiffness of CBFs in its elastic response, while providing much greater ductility and avoiding problems of buckling and irreversible yielding which affect CBFs in their post-yield phase. Arrangements such as (a) and (b) in Fig. 5.4 also have architectural advantages in allowing more space for circulation between bracing members than their concentrically braced equivalent.

EBFs have been under development for 30 years, and there are extensive design rules in seismic codes, including Eurocode 8 and US steel codes. An alternative arrangement with similar characteristics is the knee-braced frame (Fig. 5.5). The yielding element here is the 'knee brace', which remains elastic and stiff during

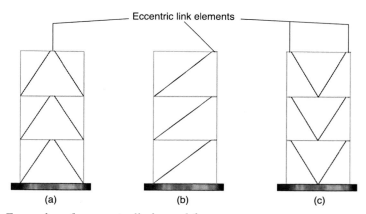

Fig. 5.4 Examples of eccentrically braced frames

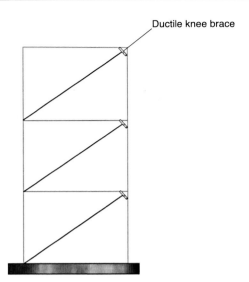

Fig. 5.5 Knee-braced frame

moderate earthquakes, but yields to provide ductility and protection from buckling in extreme events. Unlike the link in the EBF, the knee brace does not form
part of the main structural frame, and could be removed and replaced if it is
damaged in an earthquake. The concept is still undergoing development (Clément
et al. 2002) and does not yet appear in seismic codes.

5.4.6   Shear walls

(a) General

Shear walls are more rationally known as 'structural walls' in New Zealand, since
their flexural behaviour is usually more important than their shear behaviour.
Their favourable features are the provision of strength and stiffness at low cost.
The discussion below concentrates on reinforced concrete shear walls, although
plywood shear walls are widely used as an efficient bracing system in low-rise
timber-framed housing, particularly in California, and steel shear walls have
occasionally been used.

The behaviour of concrete shear walls in earthquakes has generally been excellent; they are not prone to the 'pancake' collapses which can flatten frames, and
prove so lethal to their occupants. The shear wall can be thought of as the ultimate
'strong column' which prevents formation of a soft storey. Moreover, shear walls
avoid the stress concentrations found at the beam–column joint regions of reinforced concrete frames, and avoid some of the dependence on good formwork
and steel-fixing skills associated with frames. Considerable ductility is possible in
slender shear walls which reach their ultimate strength in flexure before shear.
Stocky shear walls may be harder to make ductile, but their large potential strength
reduces the need for ductility. Offsetting these advantages to some extent, lateral
load resistance in shear wall buildings is usually concentrated on a few walls

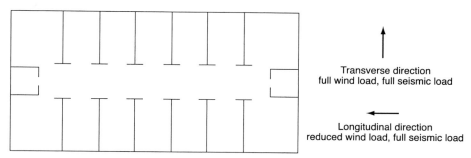

Fig. 5.6 Cross-wall construction

rather than on a large number of columns. This implies lower redundancy and possible foundation problems, including those of uplift.

Concrete shear walls often form the access cores of a building carrying lifts, staircases and service ducts. These can be readily employed as seismic-resisting elements, but the stiffness needs to be balanced on plan to prevent torsional problems arising from eccentricity between centres of stiffness and mass.

Regular cross-walls are also often found in rectangular buildings between office spaces or hotel bedrooms (Fig. 5.6); often this provides adequate seismic strength in the transverse direction of the building (where it is needed most for wind loadings) but inadequate strength in the longitudinal direction (where the wind loads are much less, but the seismic loads are similar). Another potential danger of the arrangement in Fig. 5.6 is that the partition walls are needed on upper storeys, but are discontinued at ground floor for architectural reasons, creating a potentially lethal soft storey. Shear walls at other than service cores and partition walls present barriers which may interfere with architectural and services requirements.

Shear walls on their own are a highly suitable solution for medium-rise buildings up to about 20 storeys. In taller buildings, it is likely that they need to be combined with frames to provide sufficient overall stability and stiffness.

(b) Single or isolated shear walls
The aspect ratio of a shear wall (the ratio of its height to width in the plane of loading) should normally be restricted to about 7; higher ratios may result in inadequate stiffness, problems in anchoring the tension side of the shear wall base, and possibly significant amplifications due to P–delta effects.

Aspect ratios below about 2 mark the transition from 'slender' to 'stocky' behaviour, and walls with such dimensions require considerable care in design if a ductile failure mode is required. Without this care, stocky shear walls are likely to fail in brittle failure modes such as diagonal tension or sliding shear, rather than undergoing the more ductile flexural failure possible in slender walls. Stocky shear walls may need increased strength or special detailing, including diagonal steel, to overcome these problems.

(c) Large panel precast wall systems
Large panel systems have been extensively used to provide rapid construction of medium-rise housing in seismic areas, particularly in the Balkan region and the

Former Soviet Union. In contrast with the disastrous performance of precast frames in the 1988 Armenian earthquake, panel housing performed quite well both in that event and the 1978 Bucharest earthquake. It appears that deficiencies in construction quality and lack of ductility were more than compensated for by high strength.

'Tilt-up' construction is a form of precasting extensively used in seismic areas of the USA and elsewhere for low-cost, one- and two-storey industrial sheds. The wall panels are cast horizontally on the ground at site and then 'tilted up' when they have achieved sufficient strength. They have been prone to fail in earthquakes at their connections with the roof; provided adequate strength is supplied at this connection, the system can perform satisfactorily.

(d) Frame-wall or dual systems

Combinations of moment-resisting frames with shear walls are known as frame-wall or dual systems. This combination can be structurally efficient and is favoured in both US and Japanese practice as providing good redundancy. One advantage of frame-wall systems is that the shear wall can be used to prevent a 'weak storey' forming in the moment-resisting frame. This means that the relative strength requirements to ensure a 'strong column/weak beam' frame may theoretically be relaxed. This gives more freedom in selecting beam and column sizes and there is less concern about the strengthening effect that floor slabs have on beams. Eurocode 8 and New Zealand codes allow for this, though it is not recognised by IBC (ICC 2003).

A common application of frame-wall systems is in medium- to high-rise buildings, where perimeter frames are used in conjunction with central shear wall cores. In buildings of over 50 storeys in which wind-induced motions must be controlled, 'outriggers' between the core and perimeter frame are often used (Fig. 5.7) to increase stiffness. In structures which require earthquake resistance, careful consideration must be given to the capacity design implications of using outriggers. Good ductility requires that yield occurs first in ductile modes and it must be ensured that the outriggers do not force a brittle mode, such as crushing or buckling in the perimeter columns connected to the outriggers. A possible solution is to design the outriggers to yield in a ductile manner at a load less than that corresponding to brittle failure of the columns. There appears to be no field or laboratory evidence on the efficacy of this solution.

(e) Coupled shear walls

Coupled shear walls consist of two or more walls linked by horizontal coupling beams (Fig. 5.8). The beams are often formed as a result of openings required through the wall at each floor level; the resulting structure becomes effectively a frame with very strong columns and weak beams. Most of the yielding is therefore confined to the coupling beams; provided they are adequately designed, which often involves use of diagonal steel, excellent ductility can be obtained, accompanied by good stiffness. Redundancy is also good, in that plastic energy dissipation (with the attendant risk of failure) is distributed between a number of coupling beams. It should be noted that Eurocode 8 (CEN 2004) observes that slabs are ineffective as coupling elements between pairs of shear walls, and should not be used as such.

Outrigger beams   Perimeter frame

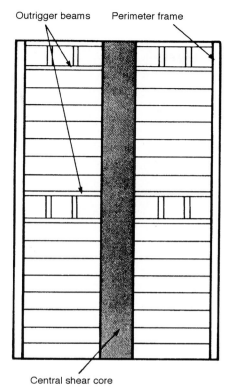

Central shear core

Fig. 5.7 Outriggers in a shear core/perimeter frame building

Limiting overall aspect ratios of coupled shear walls are similar to those for a similar unperforated single shear wall. Satisfactory efficiency of coupling beams is defined by Eurocode 8 to occur when the proportion of base moment resisted by push–pull axial forces in the shear walls is at least three-quarters of the total base overturning moment.

Coupled shear walls have been used for medium-rise construction in New Zealand, where they have been extensively researched, but appear little used for seismic resistance elsewhere. There is little field evidence as to whether their theoretical appeal translates in practice into superior performance during earthquakes.

5.4.7   Special methods of improving earthquake resistance

(a) Overview

During the past 20 years, special systems have been developed to supplement the earthquake-resisting characteristics of conventional structures by modifying their dynamic characteristics. This has generally been done by either 'passive' or 'active' devices. The passive devices either change the period of the structure or increase its damping or, more usually, do both in combination. More recently, computer-controlled 'active control' devices have been developed, which modify the structure continuously during the course of an earthquake. Warburton

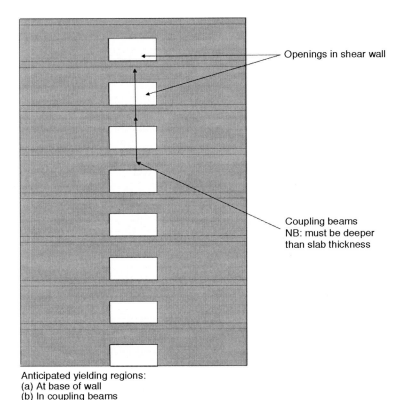

Anticipated yielding regions:
(a) At base of wall
(b) In coupling beams

Fig. 5.8 Coupled shear walls

(1992) provides a comprehensive mathematical comparison of all these methods of seismic response control.

Chapter 13 discusses the detailed design of buildings which employ these methods. The rest of this section is confined to a discussion of the general factors influencing the decision to base-isolate buildings.

(b) Seismic isolation in buildings

Seismic isolation involves mounting a building on bearings of low lateral stiffness. Laminated natural or synthetic rubber bearings are the most common form, with a typical plan dimension of 600 mm and thickness of 150–250 mm, although sliding bearings are also widely used. A typical vertical load capacity per bearing is 60 t. The intention is to increase the natural period of the building to take it away from resonance with the forcing motions of the earthquake. The bearings, because they experience high cyclic strains, also provide suitable locations for introducing hysteretic, viscous or frictional damping elements, to reduce response still further.

Seismic isolation is therefore most suitable for low- to medium-rise buildings on relatively stiff soil sites. A building period of around 1 s is the upper bound, implying a maximum height of 12–15 storeys in shear wall structures and about 10 storeys in frame buildings. Taller buildings are not suitable, partly because their period is likely already to be well away from resonance and partly for the practical reason that overturning forces would result in large uplifts on the bearings, for which it may be difficult to design. Sites with deep soft soil deposits, with a site period exceeding 1.5 s, are also not suitable, because the long-period earthquake motions associated with them mean that a shortening, not a lengthening, of the building period is needed. Where wind loads exceed 10% of the building weight, the advantages of isolation diminish considerably, although such a large percentage is very unlikely in concrete buildings. Base isolation can reduce design forces on the superstructure by a factor of up to 2 or 3. Just as importantly, by filtering out high-frequency accelerations and limiting the storey drifts (interstorey deflections), it can increase protection to non-structural elements very significantly. Moreover, occupants of base-isolated buildings are less aware of the motion caused by moderate events; this factor is valued in Japan, where perceptible earthquakes are common.

A major design problem associated with seismic isolation is the need to allow for the large horizontal deflections which would occur between the top and bottom of the bearing during a large earthquake. Services entering the building and finishing at ground level may have to accommodate deflections of the order of 100–200 mm. Flexible loops in services and suitable detailing of finishes have enabled these problems to be successfully overcome in practice.

The economic implications of seismic isolation are difficult to establish, partly because buildings are usually prototypes and so sufficiently comparable isolated and non-isolated buildings are hard to find, and partly because a conservative view has generally been taken of the reduction in superstructure forces possible because of the seismic isolation. The cost of the bearings are typically about 10% of the total structural cost, and a lower proportion of the total building cost.

The decision to isolate is usually based on the improved performance expected during an earthquake, particularly of the building contents. Where these are of high value or are needed to cope with a post-earthquake emergency, this improvement is clearly of great value.

To date, several hundred new buildings have been seismically isolated in the USA, New Zealand, Europe, Japan and elsewhere. Nuclear facilities in the UK, France and South Africa have also been isolated and its use is widespread to improve the seismic resistance of bridges. Seismic isolation has also been applied to existing buildings to improve their seismic resistance.

Seismic isolation of buildings is a relatively mature technology, which has been subject to intensive theoretical and laboratory investigation (Naeim and Kelly 1999). A few base-isolated buildings were tested to near their design specification in earthquakes in Northridge, California (1994) and Kobe, Japan (1995) and the major buildings performed well. However, codes of practice including IBC and Eurocode 8 require relatively conservative design procedures, and this may have prevented more widespread use of the technology.

5.5   Cost of providing seismic resistance

The additional cost of providing seismic resistance is hard to establish because buildings tend to be unique projects and it is difficult to compare sufficiently similar buildings which differ only in their need for seismic resistance. Indicative figures for areas of high seismicity are 20% on structural design costs, 10% on structural construction costs and significantly less on overall project costs, once building contents, services, land cost, etc. are taken into account. However, for special projects such as casualty hospitals or nuclear power related projects, the costs could be considerably greater.

An overall cost increase of less than 10% in some ways underestimates the problem of providing satisfactory seismic resistance. Catastrophic destruction of buildings, such as occurred in the Kocaeli, Turkey earthquake of 1999, has not occurred primarily as a result of penny pinching by developers or contractors, but from the absence of engineers qualified in the design and construction skill required, and the absence of checking and enforcement procedures.

These relatively low values of cost increase apply where seismic resistance is taken into account at the beginning of a project. The cost of trying to add in seismic resistance at a late stage in design is likely to be much greater. Providing seismic resistance to inadequate existing buildings is even more expensive, and can often exceed 60% of the replacement cost.

5.6 References

Arnold C. & Reitherman R. (1982). Building configuration and seismic design. Wiley, New York.
CEN (2004). EN1998-1: 2004. *Design of structures for earthquake resistance*. Part 1: General rules, seismic actions and rules for buildings. European Committee for Standardisation, Brussels.
Clément D., Blakeborough A. & Williams M. (2002). Improving the seismic resistance of steel frames using knee bracing. *Proceedings of the 12th European Conference on Earthquake Engineering*. Elsevier, Oxford.
ICC (2003). IBC: 2003. *International Building Code*. International Code Council, Falls Church, VA.
Naeim F. & Kelly J. (1999). *Design of Seismic Isolated Structures*. Wiley, New York.
Warburton G. (1992). *Reduction of Vibrations*. Third Mallet-Milne Lecture. Wiley, New York.
Wyllie L. & Filson J. (eds) (1989). Armenian Earthquake Report. *Earthquake Spectra* special supplement. Earthquake Engineering Research Institute, Oakland, CA.

6 Seismic codes of practice

'I consider that codes of practice have stultified the engineering profession, and I wonder whether an engineer can now act professionally, i.e. use his judgement.'

Francis Walley writing in
The Structural Engineer, February 2001

This chapter covers the following topics.

- The development and philosophy of codes
- Outline of code requirements for analysis, strength, deflection, detailing, foundation design and non-structural elements

6.1 Role of seismic codes in design

In most actively seismic areas, building construction is subject to a legally enforceable code which establishes minimum requirements. Even where this is not so, common practice or contractual requirements will require compliance with a code; for example, US seismic codes have very often been used in seismic areas outside the USA in the past, and the same is likely to apply to the use of Eurocode 8 in the future. In consequence, part of the normal design process will be to ensure that a set of minimum code-based acceptance criteria have been met. As with any other part of the design process, however, use of codes should not be a substitute for use of sound engineering judgement. Codes describe minimum rules for standard conditions and cannot cover every eventuality. Buildings respond to ground shaking in strict accordance with the laws of physics, not in accordance with rules laid down by a (sometimes fallible) code-drafting committee.

It should further be remembered that seismic safety results not only from the use of appropriate codes being applied with understanding in the design office but also from the resulting designs being implemented correctly on site. Disasters can happen even where reliable seismic codes of practice have statutory force; the great destruction in Western Turkey in 1999 is an often quoted example.

Seismic codes are essential tools for seismic designers; at best, they are repositories of current state of practice based on decades of experience and research. They can however limit the designer in ways that do not necessarily improve seismic safety; several examples were quoted in the previous chapter. And they form only one part of the design process.

6.2 Development of codes

Early codes were based directly on the practical lessons learned from earthquakes, relating primarily to types of construction. In 1909 following the Messina earthquake, which caused 160 000 deaths, an Italian commission recommended the use of lateral forces equal to $\frac{1}{12}$ of the weight supported. This was later increased to $\frac{1}{8}$ for the ground storey. The concept of lateral forces also became accepted in Japan although there was a division of opinion on the merits of rigidity as opposed to flexibility. After the 1923 Tokyo earthquake a lateral force factor of $\frac{1}{10}$ was recommended and a 33 m height limit imposed. In California, lateral force requirements were not adopted by statute until after the 1933 Long Beach earthquake.

After 1933, the use of lateral forces in design became widely used, with the value of the coefficients being based almost entirely on experience of earthquake damage. In 1943 the City of Los Angeles related lateral forces to the principal vibration period of the building and varied the coefficient through the height of multi-storey buildings. By 1948 information on strong motion and its frequency distribution was available and the Structural Engineers Association of California recommended the use of a base shear related to the fundamental period of the building. Once information was available on the response spectra of earthquake ground motion, the arguments over flexibility versus rigidity could be resolved. The flexible structure was subjected to lower dynamic forces but was usually weaker and suffered larger displacements.

The next important step grew out of advances in the study of the dynamic response of structures. This led to the base shear being distributed through the height of the building according to the mode shape of the fundamental mode, as originally proposed in the 1960 Structural Engineers Association of California code.

At this stage lateral forces had undergone a quiet revolution from an arbitrary set of forces based on earthquake damage studies to a set of forces which, applied as static loads, would reproduce approximately the peak dynamic response of the structure to the design earthquake. This, however, is not quite the end of the matter for lateral loads, because structural response to strong earthquakes involves yielding of the structure so that the response is inelastic.

As discussed in subsection 3.2.9, much larger design forces are required for an elastic structure without ductility than for one that can tolerate substantial plastic deformations. Because it is found in practice that the increased cost of elastic design requirements is unacceptably large, it is almost universally accepted that ductile design should apply for major earthquakes. Exceptions to this are made for structures of special importance, or where the consequences of damage are unacceptable. Although modern codes contain much useful guidance on other matters, it is the calculation of lateral design forces and the means of providing sufficient ductility that constitute, in practice, the two most vital elements for the structural engineer.

6.3 Philosophy of design

For many years, seismic codes in the West have stated or implied performance goals similar to the following (taken from SEAOC 1990).

'Structures designed in accordance with these Recommendations should, in general, be able to:

(*a*) Resist a minor level of earthquake motion without damage
(*b*) Resist a moderate level of earthquake motion without structural damage, but possibly experience some non-structural damage
(*c*) Resist a major level of earthquake motion, having an intensity equal to the strongest either experienced or forecast for the building site, without collapse, but possibly with some structural as well as non-structural damage.'

This three-tier performance standard accepts the possibility that considerable damage is likely to occur in a rare, extreme earthquake, which might not be repairable. The acceptance of survival as the aim in an extreme earthquake means that the design objective becomes that of preserving the lives of the buildings' occupants, rather than preserving property. By the end of the twentieth century, it was widely accepted that Western codes gave fairly reliable protection against life-threatening damage. However, earthquakes in the 1990s caused very large property losses in California and Japan, and the protection against preventing damage in more frequent events was considered much less reliable. Since (unlike Japanese codes) seismic codes in the West have generally considered explicitly one level of seismic load, this shortfall was perhaps not surprising. Current developments in the USA and elsewhere (see for example Fajfar and Krawinkler 1997) have been towards more explicit consideration of different levels of earthquake intensity, with performance standards varying according to their probability of occurrence. Much of the work has been in developing the 'displacement-based' methods described in subsection 3.4.3. However, Eurocode 8 (CEN 2004) and IBC (ICC 2003) still rely primarily on a single 'ultimate' limit state check as their main basis for design.

In Japan, a two-stage check has been required since the early 1980s. The structure is designed to survive a 'first-phase' event, which has a low but non-negligible probability of occurring during the building lifetime; yield strains may approach but not exceed their elastic limit based on an elastic analysis. The structure is then checked for ability to survive the 'second-phase event', which is roughly equivalent to the maximum recorded earthquake. For low-rise buildings, this involves checking a set of 'deemed to satisfy' rules (i.e. simple rules not requiring a detailed analysis), while for buildings taller than 61 m, non-linear dynamic analysis is required.

6.4   Code requirements for analysis

This section, and the ones that follow, concentrate on the European code Eurocode 8 (CEN 2004) and the US code IBC (ICC 2003), although many other codes have very similar requirements. They are intended to give a brief introduction to the codes, but are no substitute for study of the codes themselves, which have become far more complex documents than at the time of publication of the first edition of this book in 1986. Di Julio (2001) provides a helpful guide to the 2000 edition of IBC and to its predecessor UBC: 97, and Hamburger

(2003) gives a broad introduction to US seismic codes. Fardis *et al.* (2005) provide comprehensive guidance to Parts 1 and 5 of Eurocode 8. Booth and Skipp (2004) provide a background to Eurocode 8, particularly in the context of its use by UK-based engineers.

6.4.1   Equivalent static design

Most codes specify a procedure, whereby a minimum lateral strength is calculated, and then applied to the structure as a set of equivalent static forces applied up the height of the building. This is permitted for low-rise buildings without significant structural irregularities; more complex analyses are required in other cases.

The lateral strength requirement is calculated as a function of the following parameters.

(a) Building mass

This is calculated as the structural mass arising from the dead load, plus a proportion of the variable mass arising from the live load. Eurocode 8 typically specifies that 30% of the live load in office and residential loading should be included, but this might fall to 0% of the snow load in areas where snow is relatively rare and rise to up to 100% of live load for warehouses and archive buildings, and for permanent equipment.

(b) Basic seismicity

In Eurocode 8, this is expressed as the design peak ground acceleration a_g expected on rock sites, for a return period of 475 years. This is the equivalent of the Z factor in earlier US codes; IBC now provides maps for the USA of spectral accelerations expected on rock at 0.2 s and 1.0 s periods, for the 'maximum considered earthquake' with a return period of around 2500 years. In the basic CEN version of Eurocode 8, no maps are provided, and countries adopting the code are expected to provide the design accelerations in 'national annexes' to be published in conjunction with the national edition of the code.

(c) Earthquake type

In Eurocode 8, two types of site are recognised, one dominated by large-magnitude earthquakes, the other by smaller-magnitude but closer events. Different design response spectrum shapes apply to each, and the 'national annex' is supposed to specify which one should be used for a particular region. IBC does not have this explicit distinction, but adjusts the shape of design spectrum by varying the relative values of the 0.2 s and 1.0 s spectral accelerations in the seismic hazard maps of the USA referred to above.

(d) Site classification

The basic information on seismicity is presented for rock sites, but the soils overlying rock can make an enormous difference to earthquake intensity. In both Eurocode 8 and IBC, sites must be classified into one of several categories, ranging from rock to very soft soils, although the exact descriptions of the site categories vary between the two codes. In Eurocode 8, the site classification determines a factor called S, which modifies the values obtained from rock sites independently of the zone factor. In IBC, the site classification together with the basic seismicity both determine the modifications arising from soil; unlike Eurocode 8, the tendency of amplifications to reduce with intensity of earthquake is included (see subsection 4.4.1).

(e) Building function

Some buildings, such as emergency hospitals, may have a need for enhanced protection during an earthquake. IBC allows for this with an importance factor I_E, which varies between 1 and 1.5. The structural factor R (see (f) below) is divided by I_E, so effectively design forces are increased directly in proportion to I_E. In Eurocode 8, a similar factor γ_I is used to multiply the design ground acceleration on rock, so the effect is essentially the same. In Eurocode 8, recommended values of γ_I vary between 0.8 for agricultural buildings without permanent occupancy to 1.4 for emergency hospitals.

(f) Structural factor

This allows for the inherent ductility of the structure, and also the fact that during the peak transient loading of an earthquake it is acceptable to utilise more of the 'overstrength' inherent in most structures (i.e. the ratio between ultimate lateral strength and nominal design strength) than would be the case for permanent loads or wind loads. In IBC, the structural factor is called R and is a straight divisor on the required strength. R factors range from 8 for specially designed and detailed ductile frames to $1\frac{1}{4}$ for low ductility systems. In Eurocode 8, the structural factor is called q, and for medium to long period buildings also acts as a simple divisor on required lateral forces. For very short period buildings, however, the reduction due to q is limited (see Table 3.2). q factors range from 8 for very ductile structures to 1.5 for structures without seismic detailing. Unlike R in IBC, q in Eurocode 8 reduces where significant structural irregularity exists, and also depends explicitly on the 'overstrength' (ultimate lateral strength divided by lateral strength at first yield) which arises from a redistribution of forces after plastic yielding. In both IBC and Eurocode 8 the use of low-ductility structural types is restricted to areas of low seismicity; in IBC, the restrictions are more extensive than in the Eurocode, and also limit place height limits on certain structural forms.

(g) Building period

In practically every modern seismic code, the required lateral strength varies with the fundamental period of the building. This can either be assessed directly from the mass and stiffness of the structure, usually using a computer program, or from empirical formulae based on building height and structural form. IBC recognises that the latter account for the stiffening effect of non-structural elements such as cladding in lowering structural period, which will usually result in an increase in seismic load. IBC therefore restricts the advantage that can be gained from using a lower period based on a 'direct' analysis, which will generally ignore these stiffening effects. Eurocode 8 has no similar restriction.

The lateral strength requirement calculated from these procedures is then equal to the design shear at the base of the building, and it is often called the 'seismic base shear'. In order to assess strength requirements in other parts of the building, the base shear must be distributed up the height of the building. A commonly adopted formula assumes that the fundamental mode of the building is a straight line, leading to

$$F_i = F_b \frac{z_i m_i}{\sum z_j m_j} \tag{6.1}$$

where F_i is the force at level i, F_b is the seismic base shear, m_i and z_i are the mass and height at level i, and the summation $\sum m_j z_j$ is carried out for all masses from the effective base to the top of the building.

Recognising that relatively greater seismic loads may occur at the top of tall buildings due to higher mode effects, IBC modifies this formula slightly to

$$F_i = F_b \frac{(z_i m_i)^k}{\sum (z_j m_j)^k} \tag{6.2}$$

where k equals 1 for building periods less than 0.5 s (i.e. retain equation (6.1)), and k equals 2 for periods exceeding 2.5 s, with a linear interpolation for intermediate periods. Eurocode 8 allows equation (6.1) where the fundamental mode shape is approximately linear, otherwise requiring z_i and z_j to be replaced by the mode shape of the fundamental building mode.

The horizontal forces must also be distributed in plan at each level. Since the seismic forces arise from inertia effects, F_i is distributed in proportion to the mass at that level. However, special allowance is usually made for torsional effects (see subsection 3.2.7); procedures vary between codes.

This gives sufficient information to calculate not only the shears but also the bending moments at each level in the building. Recognising that higher modes produce lower bending moments at the base of a building than equivalent static analysis might suggest, IBC allows a reduction in the overturning moment at the foundation for buildings higher than 10 storeys. The foundation moment is based on 100% of the equivalent static forces from equation (6.2) for the top ten storeys, 80% of these forces for storeys above 20, with a linear interpolation on the factor for storeys between 10 and 20 above foundation level. These reductions are not permitted in Eurocode 8.

6.4.2 Response spectrum analysis

Both Eurocode 8 and IBC allow the response spectrum used as the basis for equivalent static design to be used to carry out a response spectrum analysis, and both make this mandatory for tall buildings or ones with significant structural irregularities. There are differences, however. In Eurocode 8, the results from the response spectrum analysis can be used directly. In IBC, the procedure is more complex. An equivalent static analysis must first be carried out to determine the base shear V corresponding to the period T determined from a structural analysis, and also a modified value of base shear V' corresponding to a period $c_u T_a$, where c_u ranges from 1.7 for low buildings to 1.4 for tall buildings and T_a is the period determined from an empirical equation. The total base shear from response spectrum analysis must then be adjusted to equal at least $0.85V'$ or V, if greater.

6.4.3   Time-history analysis

Linear or non-linear time-history analysis is referred to in both Eurocode 8 and IBC. A major issue is the selection of appropriate time histories. Eurocode 8 requires at least three time histories to be used, which on average match the specified design peak ground acceleration a_g and the average 5% damped spectral values must also be within 90% of the design response spectrum for the appropriate ground conditions. Either artificially generated time histories may be used, or real time histories with appropriate seismological characteristics (i.e. magnitude, distance, soil type, etc.). IBC similarly permits a minimum of three sets of artificial or real time histories, each set consisting of a pair of horizontal motions in two orthogonal directions. The average spectral values must match the design spectrum between 0.2 and 1.5 s periods.

In both Eurocode 8 and IBC, where three time histories are used, the maximum response value from the three separate analyses conducted must be used, but if seven time histories are used, the average value may be taken.

6.4.4   Non-linear static analysis

Eurocode 8 permits this type of non-linear static (pushover) analysis (subsection 3.4.3) for the following purposes in buildings.

(a) To verify or establish the 'overstrength' ratios (ultimate lateral strength divided by lateral strength at first yield) which is used in the calculation of the structural or behaviour factor q (subsection 6.4.1(f)).
(b) To estimate where plastic deformations will occur, and in what form.
(c) To assess the performance of existing or strengthened buildings, when using Eurocode 8 Part 3.
(d) To design new buildings as an alternative to the standard procedures based on scaling the results of elastic analysis by the behaviour factor q.

A detailed procedure is provided in Eurocode 8 for carrying out a non-linear static analysis, which includes special rules for torsionally eccentric buildings.

As noted above, non-linear static analysis is permitted for the design of new buildings, but designers are left without a great deal of guidance on what the maximum permissible plastic excursions should be in the design earthquake, for example the maximum rotation of a plastic hinge in a steel or concrete beam or the maximum plastic extension of steel braces. Eurocode 8 Part 1 provides information on maximum permissible curvatures in reinforced concrete beams, which can be related to plastic hinge rotations if an equivalent plastic hinge length is assumed. However, Eurocode 8 Part 1 gives no such clues for steel members in new buildings, although such information is provided in an informative annex to Eurocode 8 Part 3 for existing buildings. The need to calculate plastic capacity supply also applies of course to non-linear time-history analysis.

IBC does not currently refer to non-linear static procedures. However, the American Society of Civil Engineers Prestandard FEMA 356 (ASCE 2000) does set out detailed procedures for carrying out such an analysis for existing or retrofitted buildings. FEMA 356 provides direct information on permissible plastic rotations in steel and concrete elements (including concrete shear walls) and plastic deformations in steel tension and compression braces. As noted in subsection 3.4.3, these permissible plastic deformations are related to the performance goals 'immediate occupancy', 'life safety' and 'collapse prevention'. These values could be used to assess the values found from a non-linear static or time-history analysis.

6.5   Code requirements for strength

In general, both Eurocode 8 and IBC specify the same design strength for resisting seismic loads as for gravity, wind or other types of load. There are however important exceptions. First, in both Eurocode 8 and IBC, an important step is to check that non-ductile elements have sufficient strength so that their capacity is never exceeded. This can be achieved by the use of the capacity design principles outlined in section 3.5. Second, in IBC the concrete contribution to shear strength is usually ignored, since this tends to degrade under cyclic loading, although this only applies to high-ductility (DCH) concrete structures in Eurocode 8.

6.6   Code requirements for deflection

Storey drifts (the difference in horizontal deflection between the top and bottom of any storey) must be checked and compared with specified limits in both codes, principally to limit damage to non-structural elements. Under 475-year return events, IBC sets the maximum drift for normal buildings at between 0.7% and 2.5% of storey height, while Eurocode 8 specifies between 1% and 1.5%. P–delta effects (subsection 3.2.8) and separations between structures to prevent pounding must also be checked. Specific elements such as external cladding and columns sized for vertical loads but not seismically detailed must also be checked to confirm that they can withstand the deflections imposed on them during the design earthquake.

The calculation of deflections in Eurocode 8 follows directly equation (3.13) – that is, the elastic deflection corresponding to the design seismic forces is multiplied by the structural or behaviour factor q. In IBC, there is a similar requirement, but the multiplier instead of being the structural factor R is the 'deflection amplification factor' C_d, which is generally lower than R.

6.7   Load combinations

The seismic load combinations required by Eurocode 8 and IBC can be summarised as follows.

In Eurocode 8, the 'design action effect' (i.e. the ultimate load) is taken as due to the unfactored combination of dead plus earthquake loads, plus a reduced amount of variable loads, such as live or snow loads. Wind loads are never included with seismic loads; that is, ψ_2 is always taken as zero for wind loads. Using Eurocode notation, this is expressed as:

$$E_d \qquad = \sum G_{kj} + \qquad A_{Ed} \qquad + \qquad \sum \psi_{2i} Q_{ki} \qquad (6.3)$$

Design action effect   Dead   Earthquake   Reduced variable load

In IBC, essentially two load combinations must be considered, as follows (using IBC notation).

$$\text{Design load} = \quad 1.2D \quad + \quad 1.0E \quad + f_1L + f_2S \tag{6.4}$$

$$0.9D \quad + \quad 1.0E \tag{6.5}$$

$$\text{Dead} \quad \text{Earthquake} \quad \text{Live} \quad \text{Snow}$$

where f_1, the factor on live load, is between 0.5 and 1.0, while f_2, the factor on snow load, is between 0.2 and 0.7, depending on roof slope. Thus, member forces due to the unfactored earthquake load are combined either with 120% of forces due to gravity loads, and a reduced proportion of forces due to live and snow (but not wind) loads. In the second load combination, which for example may govern the design of columns or walls subject to uplift, the unfactored effects of earthquake load are combined with 90% of the dead load and no live load.

There is an additional requirement in IBC, which is not found in Eurocode 8. The earthquake load E is calculated as follows

$$E = \rho Q_E \pm 0.2 S_{DS} D \tag{6.6}$$

where ρ is a reliability factor, Q_E is the effect of horizontal seismic forces, S_{DS} is design 5% damped spectral acceleration at 0.2 s period and D is dead load.

The reliability factor ρ allows for the system redundancy; it lies between 1.0 for structures where the structure is highly redundant (i.e. the lateral strength is not greatly reduced by the loss of any single member) to a maximum of 1.5. Q_E is the seismic load calculated in accordance with the analysis procedures outlined in section 6.4.

The term $0.2S_{DS}D$ is stated to account for vertical seismic loads. S_{DS} corresponds to the spectral peak, which is typically 2.5 times the peak ground acceleration (pga). Therefore, for an area of high seismicity with a pga of 40%g, the term $0.2S_{DS}D$ represents $0.2 \times 2.5 \times 0.4$ or 20% of the dead load.

6.8   Code requirements for detailing
A large proportion of the seismic provisions of IBC and Eurocode 8 are concerned with detailing rules to ensure adequate ductility. IBC provides rules for steel, concrete, masonry and timber elements, while, in addition, Eurocode 8 provides rules for steel–concrete composite structures. There are detailed differences between the two procedures (see for example Booth *et al.* (1998) for a discussion on the rules for concrete) but many broad similarities exist.

6.9   Code requirements for foundations
Eurocode 8 states explicitly that capacity design considerations must apply to foundations; that is, they must be designed so that the intended plastic yielding can take place in the superstructure without substantial deformation occurring in the foundations. One way of showing this would be to design the foundation for the maximum forces derived from a pushover analysis. Eurocode 8 provides an alternative rule for ductile structures, whereby the foundation is designed for the load combination of equation (6.3), but with the earthquake load increased

by a factor 1.2Ω (reducing to 1.0Ω for $q \leq 3$) where Ω is the ratio of provided strength to design strength for the superstructure element most affecting the foundation forces. There are no similar capacity design rules in IBC.

Both codes give rules for seismic detailing of piled foundations and for site investigation requirements. The information given in Eurocode 8 Part 5 is more extensive than anything appearing in current US codes.

6.10   Code requirements for non-structural elements and building contents

Both IBC and Eurocode 8 provide simplified formulae for the forces required to anchor non-structural elements back to the main structure, in terms of the ground accelerations, the height of the non-structural element within the building (to allow for the increased accelerations at higher levels) and the ratio of natural period of the element to that of the building (to allow for resonance effects).

Rules are also given to check that items which are attached to more than one part of the structure (pipe runs, cladding elements, etc.) can withstand the relative deformations imposed on them. Extensive rules are given in Eurocode 8 for unreinforced masonry infill, which is not permitted in high-seismicity areas of the USA.

6.11   Other considerations

6.11.1   Combinations of forces in two horizontal directions

In buildings without significant torsional eccentricities and where lateral resistance is provided by walls or independent bracing systems in the two orthogonal directions, Eurocode 8 allows seismic forces in the two orthogonal directions to be considered separately, without combination. Otherwise, the forces due to each direction must be combined either by an SRSS (square root of the sum of the squares) combination or by taking 100% of forces due to loading in one direction with 30% in the other. The requirements of IBC are essentially the same.

6.11.2   Vertical seismic loads

Eurocode 8 requires vertical seismic loading to be considered in areas of high seismicity in the design of the following types of structural element

    (a) beams exceeding 20 m span
    (b) cantilevers beams exceeding 5 m
    (c) prestressed concrete beams
    (d) beams supporting columns
    (e) base-isolated structures.

Rules are given for the vertical response spectrum, which is independent of the soil type. Vertical and horizontal seismic effects can be combined either using an SRSS rule or a 100% + 30% + 30% rule, similar to that discussed above for the horizontal directions.

IBC requires that a vertical seismic load should be considered in all structures. This is calculated simply as a proportion of the dead load, the proportion increasing with the seismicity of the site (equation (6.6)).

6.12 References

ASCE (2000). *FEMA 356: Prestandard and commentary for the seismic rehabilitation of buildings*. Federal Emergency Management Agency, Washington DC.

Booth E. & Skipp B. (2004). Eurocode 8 and its implications for UK-based structural engineers. *The Structural Engineer*, Vol. 82, Issue 3, pp. 39–45, Institution of Structural Engineers, London.

Booth E., Kappos A. J. & Park R. (1998). A critical review of international practice on seismic design of reinforced concrete buildings. *The Structural Engineer*, Vol. 76, Issue 11, pp. 213–220.

CEN (2004). EN1998-1: 2004 (Eurocode 8). *Design of structures for earthquake resistance*. Part 1: General rules, seismic actions and rules for buildings. European Committee for Standardisation, Brussels.

Di Julio R. M. (2001). Linear static lateral force procedure. In: Naeim F. (ed.) *The Seismic Design Handbook*. Kluwer Academic, Boston, MA.

Fajfar P. & Krawinkler H. (eds) (1997). *Seismic Design Methodologies for the Next Generation of Codes*. Balkema, Rotterdam.

Fardis M., Carvalho E., Elnashai A., Faccioli E., Pinto P. & Plumier A. (2005). *Designer's guide to EN 1998-1 and 1998-5: Design provisions for earthquake resistant structures*. Thomas Telford, London.

Hamburger R. (2003). Building code provisions for seismic resistance. In: Chen W.-F. & Scawthorn C. (eds) *Earthquake Engineering Handbook*. CRC Press, Boca Raton, FA.

ICC (2003). IBC: 2003. *International Building Code*. International Code Council, Falls Church, VA.

SEAOC (1990). *Recommended Lateral Force Requirements*. Seismology Committee, Structural Engineers Association of California, Sacramento, CA.

7 Foundations

'There is no glory in foundations.'

Karl Terzaghi

This chapter covers the following topics. (Analysis and testing of soils is covered in Chapter 4.)

- Design objectives and capacity design
- Bearing foundations
- Piled foundations
- Deep basements
- Retaining walls
- Foundations in the presence of liquefiable soils

7.1 Design objectives

Life-threatening collapse of structures due to foundation failure in earthquakes is comparatively rare even in the extreme circumstance where soil liquefaction occurs. This is because failure of the foundation limits the amount of shaking that is transmitted into the superstructure; it is a type of uncontrolled base isolation. Foundation failure accompanied by catastrophic collapse of super-structure is therefore rare. However, foundation failures can be extremely costly; for example, liquefaction-induced failures in the port of Kobe, Japan in 1995 are estimated to have cost may billions of pounds of structural damage, with a roughly equal loss arising from economic consequences of the port's closure.

The main features to consider in the seismic design of foundations are as follows.

(a) A primary design requirement is that the soil–foundation system must be able to maintain the overall vertical and horizontal stability of the superstructure in the event of the largest credible earthquake.

(b) The foundation should be able to transmit the static and dynamic forces developed between the superstructure and soils during the design earthquake without inducing excessive movement.

(c) The possibility of soil strength being reduced during an earthquake needs to be considered.

(d) It is not sensible to design a perfectly detailed ductile superstructure supported by a foundation which fails before the superstructure can develop its yield capacity (section 7.2).

(*e*) Just as design of the superstructure should minimise irregularity, so irregular features in foundations need to be avoided. These include mixed foundation types under different parts of the structure and founding at different levels or on to strata of differing characteristics.

(*f*) Special measures are needed if liquefaction is a possibility (section 7.8).

(*g*) Special considerations apply to piled foundations (section 7.6).

7.2 'Capacity design' considerations for foundations

Capacity design (section 3.5) is accepted as a standard procedure for superstructures, but has been less widely adopted for foundations, and is uncommon in US practice. However, it is required by Eurocode 8 Part 5, and in the authors' opinion is just as valuable below ground as above. The next sections consider its application both to the sub-structural elements forming the foundations and to the surrounding soil. The basic principle is that the order of formation of yielding mechanisms must be determined, and the relative strength of superstructure, foundations and soil must be arranged so that the designer's intentions are realised in the event of a damaging earthquake.

7.2.1 Strength and ductility of foundation structures

The most straightforward case is where the strength of the foundation and its underlying soils is sufficient to support the actions corresponding to the ductile yielding mechanism chosen for the superstructure. In this case, the foundations can be assumed to remain elastic, even in the most severe earthquake, and special ductile detailing of the foundations is not required. Eurocode 8 provides a simplified rule for estimating the required capacity of foundations for this to occur (described previously in section 6.9), and in this case exempts foundations (except piles) from special seismic detailing requirements. Similarly, where the strength of the foundation exceeds that required to resist the forces imposed on it by a superstructure designed to remain elastic, no special foundation detailing is needed. Piles are an exception, and require special detailing under most circumstances; these aspects are discussed further in section 7.6. Similar capacity design rules are found in the New Zealand concrete code NZS 3101 (Standards New Zealand 1995), but they do not appear in US codes.

Where the elastic strength of the foundation structure is likely to be exceeded, some degree of ductile detailing is needed. Eurocode 8 allows for this possibility. However, foundation structures are usually difficult to inspect for possible damage and this should be borne in mind when considering whether to allow them to yield during a severe earthquake.

7.2.2 Soil response

In several earthquakes, structures where the foundation soils have failed have been observed to be less damaged than those where soil failure did not occur. Capacity design involving a yielding mechanism in the soil is therefore a possible strategy, with soil failure acting as a fuse to prevent damage of the superstructure. Since, in most cases, soils retain their shear strength even at large deformations, ductile response is usually possible, unless accompanied by excessive rotation and associated *P*–delta effects. A notable recent example is the design of the Rion-Antirion Bridge founded on poor soil and spanning the highly seismic Gulf of Corinth in Greece. Pecker (2004) describes how the bridge piers are designed to slide, in order to limit the superstructure response.

However, such a strategy is unusual, and requires extensive study. The difficulty arises from assessing with any certainty both the upper and lower bounds of the soil foundation capacity in seismic conditions. For the Rion-Antirion Bridge, special measures were taken to prevent a slip circle failure in the soil, and a carefully laid top surface of gravel was placed immediately under the foundations to ensure that sliding was controlled. Given current knowledge, the prudent course for occupied buildings will usually be to design to avoid soil bearing failure under code-specified seismic loads, and to keep ductile soil response in reserve to withstand more extreme events. However, a limited amount of sliding failure may be acceptable under some circumstances and is permitted by Eurocode 8; services such as gas and water pipes entering the building will need to be designed to accommodate such sliding. Permanent soil deformations under design earthquake loading are considered more generally acceptable in retaining walls, depending on the required performance of the wall after the earthquake, and are used to justify reduced design loads (section 7.7).

7.3   Safety factors for seismic design of foundations

7.3.1   Load factors

The advantage of carrying out a capacity design to ensure that foundations remain elastic is that there should be considerable confidence that these forces cannot be substantially exceeded. Therefore, load factors of unity on seismic loads are justified. Of course, to carry out a capacity design, information is needed about the actual yielding strength of the superstructure; where foundation construction starts before the superstructure design is complete, suitably conservative assumptions are essential.

Where the superstructure is designed for an essentially elastic response to the design earthquake, no reliance is placed on the formation of ductile yielding to limit response, and load factors of unity are probably still appropriate. However, the design earthquake is unlikely to be the maximum conceivable event, and it would be prudent for the designer to consider what might happen if the design forces are somewhat exceeded, to ensure that a brittle or unstable response is avoided. This can be particularly important in areas of moderate seismicity, where the 475-year return period motions often considered in design are usually considerably less than the maximum probable shaking.

7.3.2   Soil design strengths and material safety factors

Soil strength parameters for seismic design of the foundation structure should normally be those used for static design. However, strength degradation under cyclic loading may occur in both granular and cohesive soils, and specialist advice must be sought if this is a possibility. Conversely, the high rate of loading during an earthquake may lead to strength increases of up to 25% in some cohesive soils.

In European practice, checks on the acceptability of soil stresses follow limit state design principles, and the soil stresses are compared with soil strength divided by a material factor γ_M, which Eurocode 8 Part 5 proposes should be between 1.25 and 1.4 for seismic design. In US practice, ASCE 7-02 (ASCE 2002) states that for the load combination including earthquake 'soil capacities must be sufficient to resist loads at acceptable strains considering both the short duration of loading and the dynamic properties of the soil'; further advice is not provided.

7.4   Pad and strip foundations

7.4.1   Failure modes

In addition to transferring vertical loads safely into the soil, shallow foundations in the form of pads or strips must also transfer the horizontal forces and overturning moments arising during an earthquake. The associated potential failure modes in the soil and the foundation structure illustrated in Fig. 7.1 are now considered in turn.

(a) Sliding failure: Fig. 7.1(a)

Resistance to sliding in shallow footings will usually be mobilised from the shear strength of the soil interfacing with the footing. Passive resistance, even if significant, would only be mobilised at much larger deflections, and should generally be ignored, unless associated with a deep retaining structure such as a deep basement. In granular materials, the minimum vertical load which could occur concurrently with the maximum horizontal force must be considered, since this condition will minimise shear resistance. The maximum seismic uplift should be assessed as the sum of components due to overturning and vertical seismic accelerations, combined by the SRSS method as discussed for piles (section 7.6).

(b) Bearing capacity failure: Fig. 7.1(b)

Static bearing capacity can be determined from formulae which allow for the inclination and eccentricity of the applied load. Eurocode 8 Part 5 (CEN 2004) Annex F provides suitable expressions.

(c) Rotational failure (overturning): Fig. 7.1(c)

Where the soil is strong, the foundation may start to rotate before a bearing capacity failure occurs, particularly if the vertical load is small. In the case of pad foundations supporting a moment-resisting frame, such a rotation may be acceptable, since a frame with pinned column bases still retains lateral stability. However, the associated redistribution of moments would lead to increased moments at the top of the lower lift of columns, which would need to be designed for.

In contrast, an isolated cantilever shear wall is not statically stable with a pinned base. Rocking should, therefore, be prevented under design forces in most circumstances. Uplift can be prevented by provision of additional weight or by piles or anchors to resist the transient vertical loads, or by a wider foundation.

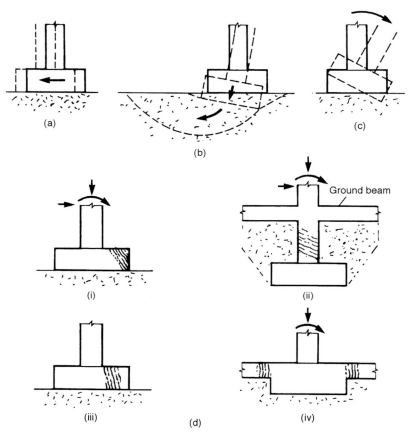

Fig. 7.1 Modes of failure in pad foundations: (a) sliding failure; (b) bearing capacity failure; (c) overturning; and (d) structural failures, where (i) shows shear failure in footing, (ii) shows shear failure in stub column, (iii) shows bending failure in footing, and (iv) shows bending failure in ground beam

(d) Structural failure in the foundation: Fig. 7.1(d)

Sufficient strength must be provided to prevent brittle failure modes in the foundation structure, such as shear failure in footings or stub columns. Ductile flexural failures may in unusual circumstances be permitted, provided the ductile detailing provisions described in previous chapters are present.

7.4.2   Ties between footings

Some form of connection is usually needed at ground level to link isolated footings supporting a moment-resisting frame. The ties prevent excessive lateral deflection in individual footings, caused by locally soft material or local differences in seismic motion. Where the footings are founded on rock or very stiff soil, however, the tendency for relative movement is much less and the ties are generally not required.

The connection can take the form of a ground beam, which will also assist in providing additional fixity to the column bases and will help to resist overturning.

Alternatively, the ground-floor slab can be specially reinforced to provide the restraint. Eurocode 8 Part 5 gives design values for the tie force which increases with seismicity, soil flexibility and axial load in the restrained columns.

7.5   Raft foundations

All of the soil failure modes illustrated in Fig. 7.1(a)–(d) may apply to raft foundations, and Fig. 7.2 shows a bearing capacity failure under a 13-storey building in the 1985 Mexico earthquake. In most cases, however, general soil failure is unlikely and the main consideration is the ability of the raft structure to distribute concentrated loads from columns or walls safely into the soil. At its simplest, the analysis would assume a uniform soil pressure distribution in equilibrium with the peak applied loads. Figure 7.3 shows that this may lead to an underestimate of shears

Fig. 7.2   Bearing capacity failure in Mexico City, 1985

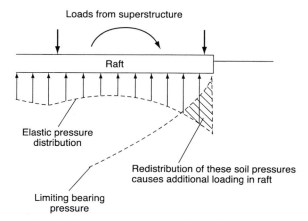

Fig. 7.3   Pressure distribution near the edge of a raft under seismic loading (after Pappin 1991)

and moments within the raft near its edge, since the soil, being poorly restrained, has low bearing capacity there. More complex analysis would allow for soil non-linearity and dynamic effects.

Partial uplift on one side of the raft under seismic overturning moments may be tolerable in the raft foundations of relatively flexible structures such as isolated tanks. Once again, however, the effect of the uplift on internal forces within the raft foundation and superstructure must be accounted for (see the discussion on tanks with uplifting bases in Priestley 1986).

7.6   Piled foundations

7.6.1   Vertical and horizontal effects

Vertical loading on pile groups during an earthquake arises from gravity loads, seismic overturning moments and vertical seismic accelerations. Since the two latter effects are not correlated, they can be combined by the SRSS method, and added to the gravity load. The procedures are straightforward, and the design of end-bearing piles is similar to that for static vertical loads. Friction piles may be less effective under earthquake conditions and require special consideration.

Horizontal response is much less easy to calculate, since the inertia loads arising from the superstructure must be combined with the effects of the soil attempting to move past the piles (Fig. 7.4). The severity of the latter effect (often called the kinematic effect) is related to the pile diameter and hence to its stiffness; flexible piles may be able to conform to the deflected soil profile without distress, but large-diameter piles are relatively much stiffer than the soil and large forces may be generated.

The most straightforward analysis for the kinematic effect assumes that the pile adopts the deflected soil profile, which may be assessed from a one-dimensional shear beam model of the soil (subsection 4.4.1). This may be overconservative, because it neglects the effect of local soil failure which will tend to reduce the curvature imposed on the pile and hence the induced moments and shears. Pappin (1991) proposes modelling the soil reaction on the piles by a series of horizontal linear or non-linear springs; the piles are modelled as vertical beams.

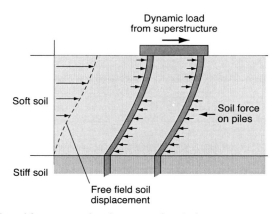

Fig. 7.4   Inertial and kinematic loading on piles (after Pappin 1991)

The deflected soil profile is then imposed on the bases of the soil springs to find the deflected shape of the piles. Programs such as SPASM (Matlock and Foo 1978) have been developed to carry out this type of analysis. The resulting actions in the piles must be combined with dynamic loading from the superstructure. It would be conservative to assume the two effects were perfectly correlated and design for a simple addition of the two effects. A dynamic analysis of the complete soil–pile–structure system would be needed for a more realistic combination.

Such analysis may indicate that plastic hinges are formed in the pile. Provided that appropriate detailing is present, plastic hinge formation may be acceptable; the detailing would take the form of closed spaced links or spirals in concrete piles, and the use of compact sections able to develop full plasticity in steel piles. However, piles are usually difficult to inspect, and their capacity to resist further lateral loads may therefore need to be assumed to have been severely reduced by an extreme earthquake. Usually, locations of plastic hinges other than at the tops of the piles are not considered acceptable. Further considerations for detailing of concrete piles are given in the next section.

Particular regions where special detailing measures may be required are as follows.

(a) The junction between pile and pile cap is a highly stressed region where large curvatures may occur in the pile. Unless adequate confinement and good connection details are present, brittle failure may occur.

(b) Junctions between soft and hard soil strata may also impose large curvatures on piles; such junctions are likely to be potential points for formation of plastic hinges.

(c) Piling through soil which may liquefy can pose special problems. In this case the pile may have a large unsupported length through the liquefied soil and should be reinforced as though it were an unsupported column. A reliable ductile behaviour will also be necessary in this situation.

General reviews of the seismic assessment of piles for design are provided by Pender (1993) and Liam Finn (2005).

7.6.2 Detailing concrete piles

Both Eurocode 8 and IBC (ICC 2003) require additional confinement steel in the form of hoops or spirals, both at the pile head and at junctions between soft and stiff soils, since these are potential plastic hinge points. Eurocode 8 also provides for minimum anchorage requirements of the vertical steel into the pilecap where tension is expected to develop in the pile.

7.6.3 Raking piles

Raking piles pose a special problem, because they tend to attract not only the entire dynamic load from the superstructure, but also the horizontal load from the soil attempting to move past the piles (Fig. 7.5), which is a particularly severe example of the kinematic interaction effect described in subsection 7.6.1. Raking piles are found to be particularly susceptible to failure in earthquakes. They should therefore be used with care in seismic regions, with particular

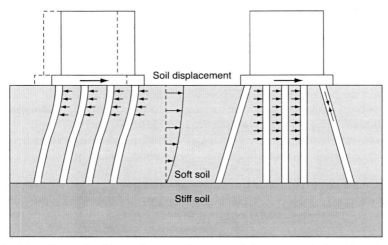

Fig. 7.5   The effect of raking piles on pile group deformation (after Pappin 1991)

attention to the kinematic interaction effects shown in Fig. 7.5, for which an explicit analysis is likely to be required.

7.7   Retaining structures

7.7.1   Introduction

During an earthquake, the soil behind a retaining structure may impose large inertia loads upon it. Where the soil is above the water table, complete collapse of the wall is unlikely to occur (Seed and Whitman 1970) but large horizontal movements are often observed. In the case of bridge abutments, these movements have led to damage or loss of support to the bridge deck in many earthquakes.

Where liquefaction can occur, complete collapse is more common; it was associated with the quay wall collapse shown in Fig. 7.6. The condition is severe because the liquefied soil imposes a hydrostatic pressure much greater than the normal active soil pressure, while the liquefaction may also weaken the restraint offered by the soil to the base of the wall.

Steedman (1998) provides a general review of methods for the seismic design of retaining walls. Useful information is also given by INA (2001).

7.7.2   Analysis of earth pressures during an earthquake

Active and passive soil pressures from granular soils on retaining walls arising from earthquakes are still commonly assessed by the Mononobe–Okabe equations, originally developed in Japan in the 1920s and quoted in Eurocode 8 Part 5 (CEN 2004). According to Steedman (1998), they are adequate for most purposes. They assume that the wall movement is large enough for an active state to develop. However, for rigid structures, such as basement walls or gravity walls founded on rock or piles, higher pressures will develop and Eurocode 8 provides a suitable equation. Expressions for seismic loads on retained clay soils are given in a Japanese standard (Anon 1980).

Fig. 7.6 Failure of a quay wall, San Antonio, Chile, 1985

Many retaining walls, for example in road cuttings or harbour walls, are designed to be able to move forward slightly during earthquake loading, either due to sliding or rotation. This results in a reduction in the required design strength of the retaining structure, in just the same way as ductility factors reduce design forces in ductile superstructures. Eurocode 8 Part 5 provides for reduction factors of between 1 and 2, depending on the circumstances, and acceptability of permanent movement. In some circumstances, significant movement is either not possible (e.g. in the retaining walls of basements in buildings) or not acceptable, in which case the full force needs to be accommodated. Steedman (1998) provides the theoretical basis for allowing for permanent movements.

7.7.3 Fluid pressures

For retaining walls with one face in contact water, the hydrodynamic interaction of the water and the wall must be accounted for. Westergaard (1933) demonstrated that, for a rigid wall retaining a water reservoir, the hydrodynamic interaction could be visualised as a portion of the water mass moving in phase with the wall. Based on his solution, Eurocode 8 Part 5 provides a simplified design pressure distribution, corrected for the wall's restraint conditions.

In addition, there is also likely to be water in the retained soil on the other face of the wall. This may move in phase with the soil, and in that case would merely add inertia to the soil. This can be accounted for by taking the total wet density of the soil in the Mononobe–Okabe equations. This is the most common situation, and Eurocode 8 Part 5 (CEN 2004) recommends that it can be assumed for soils with a coefficient of permeability of less than 5×10^{-4} m/s. However, in highly permeable soils, the water within the soil has some freedom to move independently, and will give rise to additional hydrodynamic effects. In this case, the Eurocode recommends assuming that the water is totally free; the pressures on the soil face of the wall are then the soil pressure assuming its dry density, plus an additional hydrodynamic term based on Westergaard, but reduced by the voids ratio in the soil. Steedman (1998) more conservatively recommends that walls retaining permeable soils should be checked under both assumptions – that is, either that the water is both fully restrained by the soil or that it is totally free.

7.8 Design in the presence of liquefiable soils

Two types of countermeasure are possible in the presence of liquefiable soils. Either the structures can be modified to minimise the effects of liquefaction, or the soils can be modified to reduce the risk of their liquefying.

If liquefaction is expected to be limited in extent, causing only minor local settlements, structural modification could take the form of local strengthening to cope with the settlement stresses. More radically, foundations can be moved to avoid the liquefiable soils. For example, the foundation depth can be increased to found below the levels at risk. Movement of the entire structure may also be worth considering; for example, a river bridge may be moved to a different crossing point or its span might be increased if the liquefiable material is confined to the river banks.

Foundations may also be designed which minimise the consequences of liquefaction. Possible options are as follows.

(*a*) Provision of a deep basement, so that the bearing pressures due to vertical loads are greatly reduced. Essentially, the structure is designed to float in the liquefied soil. This may be less effective in countering soil pressures due to overturning forces, and so is likely to be an option confined to relatively squat structures.

(*b*) Provision of a raft with deep upstands. The structure is designed to sink if liquefaction occurs until vertical equilibrium is regained. The solution may imply large settlements and again is most applicable to relatively squat structures.

(*c*) Provision of end-bearing piles founded below the liquefiable layers. Although this will counter vertical settlements due to gravity loads and overturning moments, the piles may be subject to large horizontal displacements occurring between top and bottom of the liquefying soil layer, and the piles must be designed and detailed to accommodate this.

The alternative strategy is to reduce the liquefaction potential of soils. A number of methods are possible and consist of three generic types, as follows. (Further information is provided by NRC (1985).)

(1) Densification, for example by vibrocompaction, which produces a more stable configuration of the soil particles. This may not be an option for existing structures, because of the settlements induced by the process.

(2) Soil stabilisation, for example by chemical grouting, which makes the soil less likely to generate rises in porewater pressure.

(3) Provision of additional drainage, for example by provision of sand drains, which tends to reduce the rise in porewater pressure.

These tend to be expensive solutions, although experience from the 1989 Loma Prieta earthquake in California suggests they can be effective (EERI 1994).

7.9 References

Anon (1980). Japanese technical standard for ports and harbour facilities in Japan.

ASCE (2002). ASCE 7-02. *Minimum design loads for buildings and other structures.* American Society of Civil Engineers, Reston, VA.

CEN (2004). EN1998: 2004. *Design of structures for earthquake resistance.* Part 1: General rules, seismic actions and rules for buildings. Part 5: Foundations, retaining structures and geotechnical aspects. European Committee for Standardisation, Brussels.

EERI (1994). *Earthquake basics: liquefaction – what it is and what to do about it.* Earthquake Engineering Research Institute, Oakland, CA.

ICC (2003). IBC: 2003. *International Building Code.* International Code Council, Falls Church, VA.

INA (2001). *Seismic Design Guidelines for Port Structures.* Working Group No. 34 of the Maritime Navigation Commission, International Navigation Association. Balkema, Amsterdam.

Liam Finn W. D. (2005). A study of piles during earthquakes: issues of design and analysis. *Bulletin of Earthquake Engineering*, Vol. 3, No. 2. Springer, Dordrecht.

Matlock H. & Foo S. H. C. (1978). *Simulation of Lateral Pile Behavior under Earthquake Motion.* University of Texas at Austin.

NRC (1985) *Liquefaction of Soils during Earthquakes, State of the Art Review.* Committee on Earthquake Engineering, National Academy Press, Washington, DC.

Pappin J. W. (1991). Design of foundation and soil structures for seismic loading. In: O'Reilly M. P. & Brown S. F. (eds) *Cyclic Loading of Soils.* Blackie, London, pp. 306–366.

Pecker A. (2004). Design and construction of the Rion Antirion Bridge. In: Yegian M. & Kavazanjian E. (eds) *Geotechnical Engineering for Transportation Projects.* American Society of Civil Engineers Geo Institute, Geotechnical Special Publication No. 126. ASCE, Reston, VA.

Pender M. (1993). Aseismic pile foundation design analysis. *Bulletin of the New Zealand National Society for Earthquake Engineering*, Vol. 26, No. 1, pp. 49–160. Wellington, New Zealand.

Priestley M. J. N. (ed.) (1986). *Seismic design of storage tanks. Recommendations of a study group of the New Zealand National Society for Earthquake Engineering.* NZNSEE, Wellington, New Zealand.

Seed H. B. & Whitman R. V. (1970). *Design of Earth Retaining Structures.* Cornell University, New York.

Standards New Zealand (1995). NZS 3101: Parts 1 and 2: 1995. *The design of concrete structures.* Standards New Zealand, Wellington, NZ. (Note: A revised edition is expected to be published in 2006.)

Steedman R. S. (1998). Seismic design of retaining walls. *Geotechnical Engineering*, Vol. 113, pp. 12–22. Institution of Civil Engineers, London.

Westergaard H. M. (1933). Water pressures on dams during earthquakes. *Transactions ASCE*, Vol. 98, pp. 418–472.

8 Reinforced concrete design

'The art of detailing reinforced concrete components for ductility comprises the skilful combination of the two materials, one inherently brittle, the other very ductile.'

Tom Paulay. In: *Simplicity and Confidence in Seismic Design.*
John Wiley, 1993

This chapter covers the following topics.

- The behaviour of reinforced concrete under cyclic loading
- Ductility in reinforced concrete, and how to achieve it
- Material specification of concrete and reinforcing steel
- Special considerations for analysis
- Design and detailing: frames, walls and diaphragms
- Prestressed and precast concrete

8.1 Lessons from earthquake damage

The principal forms of damage in reinforced concrete elements are described in Chapter 1. Once there is a loss of integrity in structural elements, mechanisms of overall or partial collapse can occur.

The need to learn from earthquake damage studies and to apply good engineering sense and judgement based on this learning cannot be emphasised too strongly. It is far more important than any amount of computation and analysis. The common sense lessons from damage studies are as follows.

(a) All frame elements must be detailed so that they can respond to strong earthquakes in a ductile fashion. Elements that are incapable of ductile behaviour must be designed to remain elastic at ultimate load conditions.

(b) Non-ductile modes such as shear and bond failures must be avoided. This implies that the anchorage and splicing of bars should not be done in areas of high concrete stress, and a high resistance to shear should be provided.

(c) Rigid elements should be attached to the structure with ductile or flexible fixings.

(d) A high degree of structural redundancy should be provided so that as many zones of energy-absorbing ductility as possible are developed before a failure mechanism is created.

(e) Joints should be provided at discontinuities, with adequate provision for movement so that pounding of the two faces against each other is avoided.

8.2   Behaviour of reinforced concrete under cyclic loading

Reinforced concrete is composed of a number of dissimilar materials. Its complex response to dynamic cyclic loading is highly non-linear and depends on the interaction between its various parts, and in particular the concrete–steel interface. Some understanding of this behaviour is necessary for the design of concrete structures in earthquake country; for a fuller description than the outline that follows, see Fenwick (1994). Other standard texts on the seismic behaviour of reinforced concrete include Paulay and Priestley (1992) and (for a European approach) Penelis and Kappos (1997). A standard text book on the seismic design of bridges by Priestley *et al.* (1996) also has excellent sections on the subject.

8.2.1   Cyclic behaviour of reinforcement

When a reinforcing bar is yielded in tension or compression and the direction of the stress is reversed, the distinct yield point is lost and the stress–strain relationship takes the curvilinear form shown in Fig. 8.1. This change in stress–strain relationship is known as the Bauschinger effect. An important result is that the stiffness of the steel is lowered as it approaches yield, compared with the initial loading cycle, which means it is more prone to buckle in the compression cycle.

The high rates of loading which occur during earthquakes may lead to increases in initial yield stress of around 20% in mild steel, although the increase is lower in high-yield steel, and the increase has also been found to be much lower in subsequent yielding cycles. These strain rate effects in reinforcement are therefore likely to be relatively minor for seismic loading. Rate effects in concrete may be much more significant (see subsection 8.2.2).

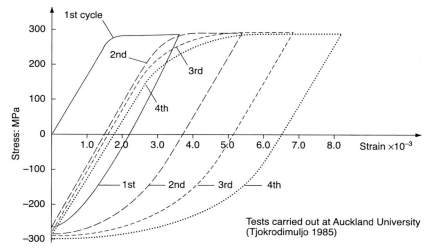

Fig. 8.1 Stress–strain relationship for mild steel reinforcement subjected to inelastic load cycles

Where the yield strength of reinforcement exceeds around 600 MPa, the margin between yield and fracture strain tends to be reduced, and the use of high-strength steel as passive (i.e. non-prestressed) reinforcement may result in structures with limited overall ductility. High-strength steel is used in prestressed concrete, discussed in subsection 8.2.10.

8.2.2 Stress–strain properties of plain concrete
Concrete on its own is weak and brittle in tension. In uniaxial compression, there is some ductility. As the concrete strength increases, the strain at maximum stress increases, but the failure tends to be more abrupt. Where there is a lateral confining pressure, the properties are very different, as discussed in the next subsection.

Strain rate effects in concrete can lead to strength increases of the order of 20% or more, and may be significant in columns with a high axial load, where flexural response is dominated by the concrete rather than the steel.

8.2.3 Stress–strain properties of confined concrete
It has long been recognised that a lateral confining pressure, when applied to concrete, can greatly increase both its compressive strength and compressive strain at fracture. Richart *et al.* (1928) showed that under triaxial loading conditions, a conservative estimate of the compressive strength $\boldsymbol{f}'_{cc}$ was given by

$$\boldsymbol{f}'_{cc} = \boldsymbol{f}'_{c} + 4.1\boldsymbol{f}_{1} \tag{8.1}$$

where $\boldsymbol{f}'_{c}$ is the cylinder strength and $\boldsymbol{f}_{1}$ is the confining pressure.

Subsequently, it was recognised that the confinement need not come from a hydrostatic pressure, but could result from the confining effect of circular or spiral reinforcement in the plane at right angles to the applied compressive stress. The mechanism is due to the tendency of concrete to expand in directions normal to an applied compressive stress. This expansion is due to Poisson's ratio effects which are enhanced (once the compressive stress reaches about 70% of the cylinder strength) by extensive microcracking. This expansion causes the confinement steel to stretch and hence develop tensile forces tending to resist the expansion. The equivalent hydrostatic confining pressure applied by the steel has an effect on strength which is quite well predicted by equation (8.1).

The effect of different quantities of confining steel on the stress–strain properties of concrete is illustrated in Fig. 8.2. It can be seen that with even small amounts of confinement, there can be a dramatic improvement in the ductility of concrete in compression, and also a significant strength increase. The confined concrete can sustain a substantial additional strain after reaching its maximum strength, and only fails when the tensile strains in the confining steel reach fracture point.

Eurocode 8 Part 2 Annex E (CEN 2004) provides equations for calculating the stress–strain curve of confined concrete as a function of the concrete strength and amount of transverse steel. The equations are based on the work of Mander *et al.* (1988). They are useful where a direct calculation is required of the rotational capacity of flexural hinges in reinforced concrete. In most cases, however, the design engineer will rely on rules for the quantity of confining steel given in codes of practice, rather than calculating them from first principles.

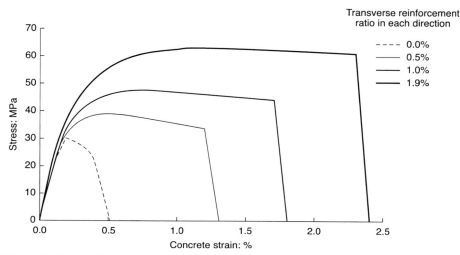

Fig. 8.2 Idealised stress–strain graph for a rectangular section with varying confinement (cylinder strength $f'_c = 30 \, MPa$)

Figure 8.2 shows typical results for a moderate-strength concrete. The mechanism by which confinement works implies that improvements are less dramatic in high-strength concrete, which is in any case more brittle in its unconfined state (ACI Committee 363 1988). This is because the effective confining pressure applied by the steel is relatively lower compared to the strength of the concrete. From equation (8.1), there is therefore a relatively lower increase in confined strength. High-strength concrete, with its high strength-to-weight ratio, may have applications in seismic design, especially for tall buildings, but it must be used with care.

Caution must also be exercised with concrete made from lightweight aggregates. In this case, the aggregates tend to crush where the confining reinforcement bears against them. The result is that the confining stress is reduced and the enhancement in strength and ultimate strain is considerably less than for normal-weight concrete of equivalent strength (Ahmad and Shah 1982).

8.2.4   Bond, anchorage and splices

During earthquake loading, reinforcing bars are subject to reverse cycle loading which in structures designed to be ductile causes the bars to yield in both tension and compression. This places a much more severe demand on the bond between concrete and reinforcement than is the case for monotonic loading. If this bond is not maintained, bars will lose their anchorage and not be able to develop the forces needed to resist earthquake effects, and they will also lose continuity at lapped splices. Bond strength under cyclic loading is improved where the concrete is confined by closely spaced hoops or spirals.

Some implications for design are as follows.

(*a*) Anchorage of bars in earthquake-resisting structures needs special attention. Bars that are in highly stressed regions should terminate in a bend or hook to provide mechanical anchorage.

(b) Anchorage and splicing of bars should be avoided in areas where plastic hinges are expected to form. One advantage of capacity design procedures (section 3.5) is that they provide the designer with some confidence in identifying non-yielding areas of the structure where anchorage and splicing may take place.

(c) The concrete where bars are anchored or spliced should be well confined with hoops or spirals.

8.2.5   Flexure and shear in beams: reversing hinges

Ductile concrete frames are designed so that plastic hinges form in the beams under design earthquake loading. The plastic hinge regions must therefore be able to sustain large plastic rotations without significant loss of flexural strength and without shear failure.

Under seismic loading, ductile moment-resisting frames with relatively low levels of gravity loading form plastic hinges at the ends of the beams. These hinges yield first in one direction and then the other as the frames sway to and fro during a large earthquake. Under these conditions of reversing load, diagonal shear cracks form in the plastic hinge region, which widen progressively with the number of loading cycles as both flexural and shear steel accumulate plastic tensile strains. This tends to destroy the contribution of aggregate interlock and dowel action to shear resistance. Under static loading conditions, this contribution can be safely included, and is accounted for in codes as the concrete contribution to shear resistance. Under seismic loading, some or all of this contribution will be lost, and codes specify that it should be discounted in beams, unless the shear stresses or ductility demands are low or a significant compressive stress is present.

The loss of 'concrete contribution' means that shear has to be resisted entirely by a truss action formed by the steel flexural and shear steel as tension members and diagonal concrete compression struts. The widening diagonal cracks then lead to another consequence; in order for the compression strut to take its load after a stress reversal, the diagonal crack across it must first close (Fig. 8.3). This leads to a situation where there is very little resistance to shear and hence stiffness around the midpoint of the loading cycle. When the effect of this shear deformation from yielding of the shear reinforcement is added to the flexural deformation from yielding of the main bars, the characteristic pinched shape of a hysteresis loop is obtained (Fig. 8.4).

Strength degradation and eventual failure in a reversing hinge can occur in a number of ways. First, the longitudinal bars may fracture in tension or they may buckle in compression; the reduction in restraint from the shear steel as the latter yields in tension increases the tendency to buckle as does softening due to the Bauschinger effect (subsection 8.2.1).

Second, the opening and closing of the diagonal cracks in the web causes loss of the concrete contribution to shear strength, as explained above. It can also lead to strength deterioration of the concrete in the diagonal compression strut, eventually leading to failure of the strut at stress levels considerably less than those that can be sustained under monotonic conditions. Failure then generally occurs close to one of the major cracks in the plastic hinge, and it is accompanied by high shear

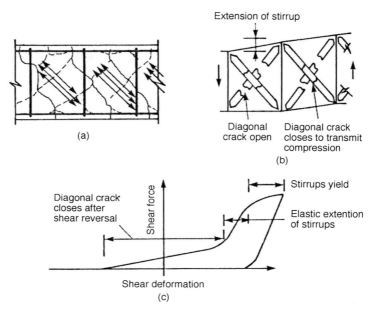

Fig. 8.3 Shear deformation in reversing hinge zones: (a) crack pattern; (b) deformation of truss; and (c) shear versus shear deformation in reversing hinge

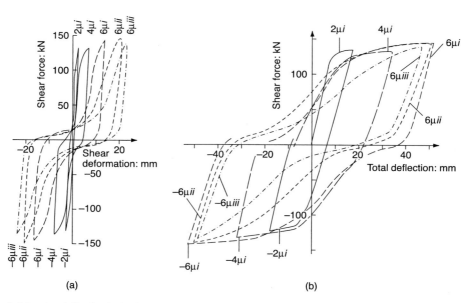

2μi denotes deflection in the first cycle of load to twice yield.
2μii denotes deflection in the second cycle of load to twice yield, etc.

Fig. 8.4 Load–deflection test results for a beam developing a reversing plastic hinge:
(a) shear force versus shear deformation; and (b) shear force versus total deflection
(i.e. shear plus flexure) (from Fenwick et al. 1981)

Fig. 8.5   Sliding shear failure in a reversing plastic hinge

displacement (Fig. 8.5) – hence the name sliding shear. Sliding shear failure can be prevented by the addition of diagonal shear steel, and Eurocode 8 (CEN 2004) requires such steel where reversing shear forces exceed a given threshold.

Finally, failure might occur due to fracture of the concrete in the flexural compression zone, particularly if there is inadequate confinement steel, but this is less likely in the absence of an overall compressive force in the beam.

Crack widths under reversing cyclic loads become progressively larger. This is because part of the yielding tension force in the main steel is resisted by the concrete, so that for equal areas of top and bottom steel there is never enough tension to force the compression steel to yield in compression and recover some of the plastic tension yielding from previous cycles. This tendency for cracks to widen is increased by aggregate particles becoming dislodged and wedging open the cracks. Tensile plastic strains in the steel therefore accumulate and the overall length of the beam increases. Even for unequal areas of top and bottom steel, the side with the most steel will tend to accumulate tensile strain. As a result, the beams elongate in a severe earthquake, and this imposes additional rotations on the lowest columns, particularly at the two ends of a frame (Fig. 8.6). Thus, plastic hinges may form at both top and bottom of the lowest columns, even where a capacity design has been carried out to ensure a 'strong column/weak beam' system. Moreover, the elongation imposes severe conditions on attached elements such as cladding panels and floor diaphragms, which have implications for the design of their connection and bearing arrangements.

Current codes do not generally require an explicit consideration of these elongation effects in ductile concrete frames, although they have been researched and can be very significant (Fenwick and Davidson (1993); Fenwick and Megget (1993); Lau *et al.* (2002); Fenwick *et al.* (2005)). They illustrate that a robust and cautious

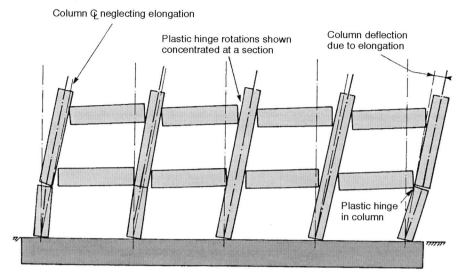

Column ℄ neglecting elongation

Plastic hinge rotations shown
concentrated at a section

Column deflection
due to elongation

Plastic hinge
in column

Fig. 8.6 Effect of beam elongation on deflections and rotations in a ductile frame

approach is needed in seismic design to matters such as designing for the possibility
of plastic hinges forming in columns and provision of anchorage and bearing for
precast floor units. Note that even quite sophisticated non-linear dynamic seismic
analysis will not generally model elongation effects and their consequences.

Some implications for design are as follows.

(a) Closely spaced transverse steel is required at potential plastic hinge points
for four reasons
  • to provide adequate shear strength
  • to limit shear deformations
  • to provide buckling restraint to main steel
  • to confine the concrete in the flexural compression zone, in order to ensure
    its integrity.
(b) The concrete contribution to shear strength in beams tends to degrade under
conditions of cyclic loading.
(c) Diagonal shear steel is needed at potential plastic hinge regions under conditions of high levels of reversing shear.
(d) Unexpected and unquantified effects may occur during severe earthquake
loading, and the design needs to be sufficiently robust to accommodate
them. In particular, elongation of beams may require special detailing of
lower columns and of restraint and bearing details to precast floors.

8.2.6 Flexure and shear in beams: unidirectional hinges
The previous section considered beams in which two plastic hinges form under
extreme earthquake loading, one at each end of the beam. With successive cycles
of earthquake loading, the plastic hinges rotate first in one direction and then
the other (Fig. 8.7(a)). This is the situation likely to apply where the seismic

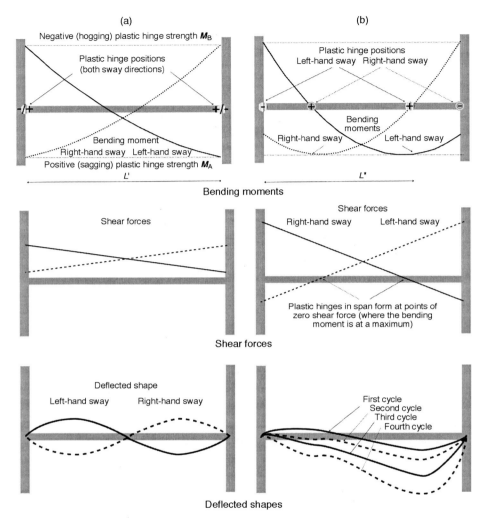

Fig. 8.7 Reversing and unidirectional hinge beams: (a) reversing hinge beam (low vertical load); and (b) unidirectional hinge beam (high vertical load)

resistance is provided by a perimeter frame which takes relatively low gravity loads, and the main vertical load-bearing system is formed from gravity-only internal frames.

A different situation occurs when the beams in a seismic frame carry significant gravity loads. In this case, the positive (sagging) moments due to the vertical loads may be sufficient to cause plastic hinges to form in the beam span (Fig. 8.7(b)). Four plastic hinges will then form during earthquake loading; a left-hand support hinge and a right-hand span hinge during one direction of loading, and a right-hand support hinge and left-hand span hinge during the other. For beams which have unvarying bending strength along the beam, it can easily be shown that unidirectional hinges will form if the shear forces at the two ends of the beam differ in sign when the first hinge forms at a support. This condition corresponds to

$$w > 2(\boldsymbol{M}_A + \boldsymbol{M}_B)/(L')^2 \tag{8.2}$$

where w is the vertical loading per unit length on the beam (gravity plus vertical seismic accelerations, assumed uniform along the beam), $\boldsymbol{M}_A$ and $\boldsymbol{M}_B$ are the positive and negative flexural strengths of the beam and L' is the clear span. Where w exceeds this limit, the distance between the span and support hinges forming in any cycle is L^*, where

$$L^* = \sqrt{2(\boldsymbol{M}_A + \boldsymbol{M}_B)/w} \tag{8.3}$$

The important consequence is that each hinge rotates in one direction only. This avoids some of the degradation effects due to hinge reversals noted previously, and the total rotation capacity under such loading (the ductility supply) can be around twice that for reversing hinges. However, there is an offsetting disadvantage: the plastic rotations increase cumulatively instead of alternating between two extremes and the ultimate rotation of unidirectional hinges may be quickly reached. Hence the overall ductility demand is likely to be much greater than for frames with reversing hinges, and becomes linked more critically to the duration of the earthquake (i.e. the number of loading cycles) as well as its maximum intensity. Elongation of the beams is also more severe.

Another consequence is that the maximum shear in the beam is likely to increase because the two plastic hinges are separated by a distance less than the clear shear span L'. For beams with unidirectional hinges, the equation in Fig. 3.27 therefore needs to replace L' by L^*, the distance between span and support hinge points.

Some implications for design are as follows.

(a) Where the beams of frames take significant vertical as well as lateral loads, unidirectional hinges may form, which increases the ductility demand. The effect is greater for large magnitude, long duration earthquakes (Fenwick *et al.* 1999).

(b) Peak shear forces in beams are likely to be greater for unidirectional than for reversing hinges.

(c) Formation of plastic hinges within the beam span – and hence the undesirable unidirectional effects described – can be prevented by placing additional bottom steel which stops short of the plastic hinge region at the beam supports.

8.2.7   Flexure and shear in columns

Column failure is likely to have more disastrous consequences than beam failure, because the loss of support extends to all floors above the failed column. Columns therefore need additional protection to guard against flexural or shear failure.

The differences between beam and column behaviour under cyclic loading arise from the compressive load that a column carries. This has two major consequences.

(1) Shear and flexural cracks opening under one cycle of loading are likely to close under the influence of the compressive load in the reverse cycle. The

pinching of the hysteretic loops found in beams (Fig. 8.4) is therefore less pronounced, and the loss of 'concrete contribution' to shear strength (subsection 8.2.5) does not occur. Codes of practice therefore allow the full shear strength under static loading conditions to be assumed where a significant compressive stress is present.

(2) The additional compressive stress increases the cyclic compressive strain that the concrete must sustain, and as a consequence the concrete strength will quickly degrade at plastic hinge locations unless adequate confinement steel is present.

8.2.8 Flexure and shear in slender shear walls

A slender shear wall is defined as one in which the height exceeds twice the width. Under these conditions, a suitably designed wall can form a ductile flexural hinge at the base which achieves a level of ductility only slightly less than that of a well-detailed frame (Fig. 8.8). In New Zealand practice, the term 'shear wall', with its connotations of brittle shear failure, is felt to be a misnomer, and the term 'structural wall' is preferred.

Note the absence of stiffness and strength degradation in Fig. 8.8 and the absence of significant pinching in the hysteresis loops. In order to achieve this ductile flexural behaviour, a number of conditions must be met.

(a) The longitudinal reinforcement in the compression region of the hinge must be restrained against buckling by closely spaced links.

(b) The compression edge of the wall must not be so slender as to suffer a buckling failure. A thickening of the edge, or the bracing provided by a transverse wall, helps prevent this.

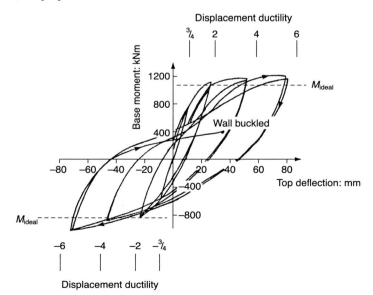

Fig. 8.8 Bending moment versus lateral displacement in a shear wall with a flexurally dominated response (from Paulay and Priestley 1992)

(*c*) The concrete must be well confined at the ends of the wall where it is required to sustain high compressive cyclic strains.

(*d*) Plastic hinge formation must occur in a location (usually the base) where there is adequate detailing to sustain large plastic deformations. In order to achieve this, both Eurocode 8 (CEN 2004) and the New Zealand concrete code (NZS 1995) stipulate a capacity design procedure which ensures that flexural yielding occurs only at the base of the wall. US practice (ACI 2002) does not include this requirement.

(*e*) The shear strength of the wall throughout its height must be sufficient to sustain the chosen plastic flexural hinge mechanism. Once again, Eurocode 8 and the New Zealand code have special requirements to achieve this which do not appear in US practice.

Shear failure in shear walls can occur in diagonal tension or compression in a similar to way to beams. Eurocode 8 (CEN 2004) provides for a strength calculation which is similar to that for columns, with an allowance for the favourable effect of compressive axial loading on the wall. ACI 318 (ACI 2002) by contrast makes no such allowance.

Another form of shear failure is sliding shear at horizontal planes; this can be resisted by shear friction across any horizontal crack and by dowel action. Distributed vertical reinforcement has several roles in this: it helps to distribute cracking, provides dowel resistance and also helps clamp concrete surfaces together. Construction joints are clearly potential sliding shear planes; they should be well roughened, cleaned of loose debris and checked for strength using a shear friction calculation.

Anchorage failure of the main reinforcement steel leads to loss of strength and must be detailed against by providing generous anchorage. Under cyclic loading, yielding of the steel will occur progressively further down the reinforcing bars, and may penetrate into the wall foundation by around 20 bar diameters for a displacement ductility of 6. Full tension anchorage of the bars is therefore required beyond this point.

8.2.9   Squat shear walls

Where the height-to-width ratio of a wall is less than about 2, the shear force necessary to develop a flexural hinge becomes relatively large, and ductile flexural behaviour may be hard to achieve. Often, this is not of concern because the inherent strength of a squat shear wall enables seismic action to be resisted without the need to develop much ductility. However, provided sliding shear failure is prevented, some ductility can be achieved through yielding of vertical reinforcement. The provision of diagonal reinforcement anchoring the base of the wall to its foundation has been shown to improve resistance to sliding shear greatly (Paulay *et al.* 1982), and is required in Eurocode 8 (CEN 2004) where high shear stresses are present.

8.2.10   Prestressed concrete

The behaviour of prestressed concrete beams has similarities with that of passively reinforced columns; thus, the prestressing force improves shear resistance and

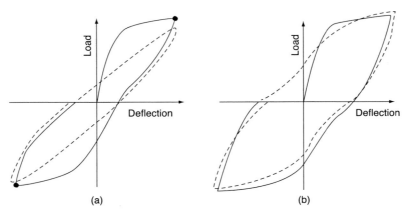

Fig. 8.9 Cyclic response of (a) prestressed and (b) reinforced concrete

reduces the tendency for stiffness degradation but increases concrete compressive strain demands, and the steel itself has a lower fracture strain, which in itself tends to reduce ductility. Figure 8.9 compares the cyclic response of prestressed and passively reinforced members; the lower ductility and hysteretic energy dissipation in the former is evident.

There is limited codified information on the seismic design of prestressed concrete buildings, and it is not covered by Eurocode 8 Part 1 (CEN 2004), although Part 2 gives limited advice for bridges. ACI 318 (ACI 2002) permits limited partial prestressing but provides little advice. However, the New Zealand code NZS 3101 (NZS 1995) has a section and commentary on the subject, while Fenwick (1994) provides further background.

8.2.11   Non-ferrous reinforcement in seismic-resisting structures

In non-seismic applications, increasing use is being made of non-ferrous reinforcement to provide tensile strength both for concrete and for non-cementitious resins. The reinforcement takes the form of fibres made of carbon, various types of plastic (aramid, polyethylene) or glass. They are characterised by high-tensile strength and good corrosion resistance, compared to steel, but possess little or no ductility. Their main use in seismic applications has been as external jacketing applied as confinement to existing concrete columns, to which they are bonded using epoxy resins; see the discussion in section 8.2 of Priestley *et al.* (1996). This confinement can provide additional shear strength and also improve the flexural ductility by increasing the strength and ultimate strain of the confined concrete. It can also enhance the bond strength between reinforcing bars. Annex A, section 4.4 of Eurocode 8 Part 3 (CEN 2004) gives information on the use of fibre-reinforced plastics as a jacketing material.

8.3   Material specification

Table 8.1 shows the material specifications of Eurocode 8 (CEN 2004) and ACI 318 (ACI 2002). The Eurocode requirements are shown for the regions of high

Table 8.1 Concrete and steel specifications for high-ductility seismic-resisting structures

	Eurocode 8 (CEN 2004)	ACI 318 (ACI 2002)
Concrete cylinder strength: Minimum Maximum	20 MPa (Note 1)	21 MPa (Note 2)
Reinforcement: General	Plain round bars are only acceptable as hoops or ties; otherwise deformed or ribbed steel must be used	
Yield strength	400–600 MPa	A706: 420 MPa A615: 300 MPa or 420 MPa
Minimum tensile strain (on 200 mm)	Strain at ultimate tensile strength Class B: ≥5% Class C: ≥7.5%	Strain at failure A706: 10–14% A615: 7–12% (depending on bar diameter)
Ultimate tensile strength: Yield or 0.2% proof strength	Class B: ≥1.08 Class C: between 1.15 and 1.35	≥1.25
Upper characteristic yield strength: Nominal yield strength	≤1.25	(See equivalent requirement below)
Actual yield strength less specified yield strength	(See equivalent requirement above)	≤21 MPa

Note 1: Concrete with cylinder strength exceeding 50 MPa is not covered by Eurocode 8.
Note 2: The cylinder strength of lightweight concrete may not normally exceed 28 MPa, but this may be increased if justified by tests.

ductility (DCH) structures expected to form plastic hinges; those for other regions and for medium ductility (DCM) structures are sometimes relaxed.

The rationale behind the main requirements of Table 8.1 is as follows. Minimum concrete strength is specified to ensure a reasonable level of strength and ultimate strain, while the reasons for restrictions on maximum strength were discussed in subsection 8.2.2.

Reinforcement with reasonable ultimate tensile strain is an obvious requirement to ensure ductility. The restrictions on the difference between actual and specified yield strength (or equivalently, between actual and upper characteristic yield strength in Eurocode 8) arise from capacity design considerations. Thus for example, the required shear strength of a beam or column should be based on the actual flexural strength that the beam achieves, and if on site this exceeds the designer's assumptions, the shear strength provided may be insufficient to develop the actual flexural strength. The minimum ratio between ultimate tensile and yield strength is to ensure that yielding in regions of rapidly changing moment (i.e. high shear) spreads over a reasonable length of the beam, thus producing good ductility (see equation (8.5)).

8.4   Analysis of reinforced concrete structures

Concrete structures may be analysed by any of the methods discussed in Chapter 3. This section discusses some particular aspects with respect to modelling.

8.4.1   Modelling the stiffness of reinforced concrete members

The stiffness of concrete structures tends to reduce under severe ground shaking. This may be partly due to the formation of plastic mechanisms, and this should be adequately accounted for in a ductility-modified response spectrum analysis or a non-linear analysis in which the plastic hinges are explicitly modelled. However, cracking of concrete is likely away from plastic hinge positions, and therefore basing member properties on the uncracked concrete section will overestimate the stiffness and underestimate the natural periods of the structure. Generally speaking, this will lead to an overestimate of seismic forces and an underestimate of deflections.

IBC (ICC 2003) requires that 'the stiffness properties of concrete (and masonry) members shall include the effect of cracked sections'. Common US practice is to take 50% of gross section properties. Seismic loads in IBC are related to an empirically determined period which includes the stiffening effect of non-structural elements such as cladding, so in principle IBC conservatively allows for the effect of cracking on increasing deflections, but does not base seismic forces directly on a potentially unsafe reduction in stiffness.

Eurocode 8 (CEN 2004) specifically allows the stiffness of concrete (and masonry) members to be based on 50% of the gross stiffness, in the absence of a more detailed analysis. Significantly, there is no requirement to relate design forces to those based on an empirically determined period, although this is available as an option.

Paulay and Priestley (1992) recommend that the fraction of gross stiffness taken into account should account for (among other things) the axial compressive load in the member, as shown in Table 8.2. The lower values should be used with some

Table 8.2 *Effective member moments of inertia (from Paulay and Priestley 1992)*

	Range	Recommended value
Rectangular beams	$0.30\text{–}0.50I_g$	$0.40I_g$
T and L beams	$0.25\text{–}0.45I_g$	$0.35I_g$
Columns		
axial load $> 0.5f'_c A_g$	$0.70\text{–}0.90I_g$	$0.80I_g$
axial load $= 0.5f'_c A_g$	$0.50\text{–}0.70I_g$	$0.60I_g$
axial load $= -0.05f'_c A_g$	$0.30\text{–}0.50I_g$	$0.40I_g$

A_g = gross area of section; I_g = moment of inertia of gross concrete section about the centroidal axis, neglecting the reinforcement; f'_c = concrete cylinder strength.

care if employed as the basis for assessing design seismic forces (as opposed to deflections).

More recently, work by Paulay and Priestley has suggested that effective stiffness should be related directly to strength (Priestley 2003, ch. 2). This appears to work well for high ratios of steel, but may underestimate the stiffness of lightly re-inforced sections because the tension stiffening effect of the concrete is neglected.

8.4.2 Damping in concrete structures

In a structure responding plastically to an earthquake, most of the damping is hysteretic and in a ductility-modified response spectrum analysis, this is represented by the ductility reduction factor (for example, the q factor in Eurocode 8 or R in IBC). Therefore, no separate allowance needs to be made for the level of viscous damping. However, in an elastically responding structure, the usual assumption of 5% modal damping may need to be adjusted. ASCE 4-98 (ASCE 1998) recommends modal damping values in reinforced concrete structures of 4% to 7%, the former being applicable where stresses are generally below half yield, and the latter where stresses are approaching yield. For prestressed members, these values reduce to 2% and 5% respectively. These values would be appropriate in an elastic response spectrum or time-history analysis.

In a non-linear time-history analysis, the hysteretic damping is accounted for explicitly, since yielding is taken directly into account. Viscous damping to account for energy dissipation in the elastic range needs to be used rather carefully, and the ASCE 4-98 values quoted above may be unconservative. This is because although they are appropriate while the structure is in its elastic range, they may substantially overestimate the dissipated energy when the structure yields, for reasons discussed in subsection 3.2.2.

8.4.3 Assessing the rotational capacity of concrete elements

In a well-designed concrete frame or shear wall structure, yielding and energy dissipation under extreme seismic loading takes place through rotation of plastic hinges forming within the structure. The ductility available then depends on the ultimate rotational capacity of those plastic hinge regions.

In conventional code design using elastic analysis, the adequacy of the plastic hinge regions is obtained by designing for a strength depending on a structural factor (e.g. the q factor in Eurocode 8 or R factor in IBC) and then applying detailing rules given in the code which correspond to the structural factor adopted.

Non-linear static (subsection 3.4.3) or time-history analysis (subsection 3.4.2) enables the local ductility demand at plastically yielding regions to be determined directly, which in principle is a much more satisfactory procedure. Where these regions are modelled as discrete plastic hinges, the analysis will result in values of plastic rotation, which must then be compared with the available capacity.

The most direct advice is given by FEMA 356 (FEMA 2000), which provides limiting rotations corresponding to different performance goals for a variety of elements (see subsection 3.4.3(e)). Eurocode 8 (CEN 2004) Part 3 also provides limiting rotations, but IBC (ICC 2003) provides no such direct advice (although this may be added to later editions).

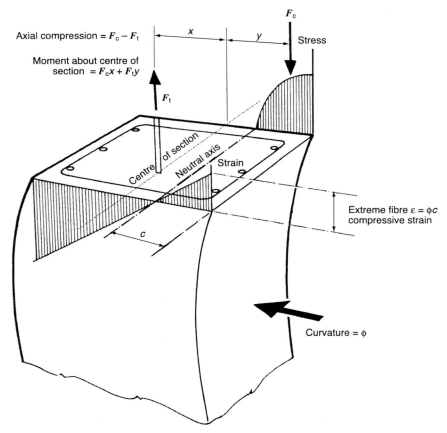

Fig. 8.10 Calculation of moment–curvature relationship under uniaxial bending, assuming plane sections remain plane

A calculation is also possible from first principles. This calculation starts by establishing the relationship between curvature of a section and moment, taking into account any axial load that may be present. With the knowledge of the stress–strain characteristics of both concrete and steel, and on the assumption that plane sections remain plane, this is in principle straightforward to do (Fig. 8.10), and many programs exist to perform the calculation (e.g. the freely downloadable code BIAX, 1992). Note that the stress–strain characteristics of concrete depend on the amount of confining steel (Fig. 8.2). Note also the limitations of the assumption of plane sections remaining plane. It is strictly only true for monolithic sections with zero shear, although the shear stress needs to be very high for serious error to occur. More importantly for reinforced concrete sections, slip between steel and concrete is not included. Therefore, under the conditions of high shear and bond slip likely in plastic hinge regions with high ductility demand, the results are approximations.

The curvature–moment relationship can be transformed to a rotation–moment relationship by multiplying the curvature by an effective plastic hinge length L_{pl}

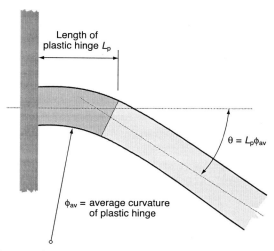

Fig. 8.11 Calculation of effective plastic hinge rotation

(Fig. 8.11). Part 3 of Eurocode 8 (CEN 2004) provides the following expression for L_{pl}

$$L_{pl} = 0.08L_v + \alpha_{sl}d_b f_y \tag{8.4}$$

where $L_v = M_u/V_u$ the bending moment to shear force ratio at the critical section of the hinge, d_b and f_y are the diameter and yield strength of the longitudinal reinforcement used in the hinge zone and the value of α_{sl} is 1 if there is slippage of the longitudinal bars from their anchorage beyond the member end, or 0 if there is no slippage. Other expressions for hinge length are also available (see Mander *et al.* 1988; Riva and Cohn 1990; Priestley *et al.* 1996).

The plastic rotation θ_p of the hinge corresponding to a curvature ϕ_p can then be estimated as

$$\theta_p = (\phi_p - \phi_y)L_{pl} \tag{8.5}$$

where ϕ_y is the curvature at first yield.

The simplest approach would then be to calculate the ultimate rotation by substituting $\phi_u = \phi_p$ in equation (8.5), where ϕ_u is the curvature at concrete or steel fracture. While this might be satisfactory if only one significant loading cycle is expected, it takes no account of the significant stiffness and strength degradation under cyclic loading typical for seismic loading. Part 3 of Eurocode 8 deals with the strengthening of existing structures and recommends that the ultimate plastic rotation of plastic hinges should be assessed from equation (8.6). A similar but less conservative equation is quoted by Otani *et al.* (2000) as appearing in the recently revised Japanese seismic provisions.

$$\theta_p = \theta_u - \theta_y = (\phi_u - \phi_y)L_{pl}(1 - 0.5L_{pl}/L_v) \tag{8.6}$$

where θ_u and θ_y are the hinge rotations at yield and ultimate and the other symbols have the same meanings as before. In order to define θ_u and θ_y more precisely, rotations should be calculated in Eurocode 8 as shown in Fig. 8.12. Yield refers

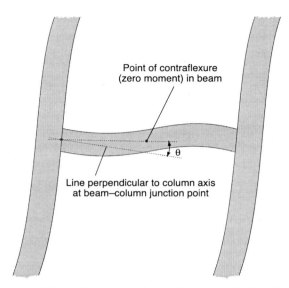

Fig. 8.12 Definition of beam hinge rotation in Eurocode 8 Part 3

to conditions when the reinforcement at the most critical section first reaches its yield stress, while the ultimate rotation corresponds to a near collapse limit state. The plastic rotation corresponding to significant damage is taken by Eurocode 8 as $\frac{3}{4}$ of this value.

Other approaches are possible for evaluating the effect of cyclic degradation on the capacity of plastic hinges. Park *et al.* (1987) provide expressions which relate hinge capacity directly to the hysteretic energy they dissipate, while programs such as DRAIN-2DX (1993) have elements making direct allowance for stiffness and strength degradation.

8.5   Design of concrete building structures

The remaining sections of this chapter discuss design considerations for concrete structures, with particular reference to the provisions of Eurocode 8 and ACI 318. For a more detailed discussion of ACI 318, refer to Derecho and Kianoush (2001) and Mo (2003).

8.6   Design levels of ductility

Eurocode 8 recognises two classes of ductility in concrete structures designed for areas of high or moderate seismicity. Ductility class 'high' (DCH) structures may be designed for lower lateral strength, but have stringent rules for detailing and strength assessment. These rules are relaxed (sometimes substantially) in ductility class 'medium' (DCM) structures, but at the expense of a lateral strength requirement which is about 50% greater. The design effort required for DCM structures is significantly less than that for DCH, as noted in the following sections. A third, 'low' ductility class (DCL) is also defined in Eurocode 8, requiring the highest lateral strength but with no special seismic rules so that design to the non-seismic Eurocode 2 for concrete suffices. However, DCL structures may only resist seismic forces in areas of low seismicity. Non-seismically detailed frames may be used in areas of high and moderate seismicity, but they can only be used for supporting gravity loads and their contribution to lateral resistance must be neglected, as discussed in subsection 8.7.6.

Similar classifications apply in US practice in ACI 318, although the ductility levels are classified as 'special', 'intermediate' and 'ordinary'. However, there is not a one-to-one correspondence with the Eurocode ductility classes. In particular, intermediate and ordinary moment frames are not permitted in areas of high seismicity, and the Eurocode DCH and DCM classifications are essentially sub-divisions of the ACI special ductility level.

8.7 Design of reinforced concrete frames

8.7.1 Introduction
Moment-resisting concrete frames rely on the rigidity of the beam–column joints to resist lateral loads, rather than on shear walls or cross-bracing. They are some-times called unbraced frames.

8.7.2 Preliminary sizing
Codes place restrictions on the range of geometries permitted in ductile frames, as shown in Table 8.3. The rationale behind the main requirements shown is as follows.

The restrictions on beam-to-column width ratios are to ensure a flow of moment between beams and columns without undue stress concentrations and to harness

Table 8.3 Code guidance on beam and column dimensions for high-ductility frames

Columns:	
ACI 318 (ACI 2002)	Shortest c/s dimension ≥305 mm; ≥0.4 perpendicular direction
Eurocode 8 (CEN 2004)	Shortest c/s dimension ≥250 mm; C/s dimension ≥ one-tenth of larger distance between point of contraflexure in column, and end of column, for bending in the plane of dimension considered (unless axial forces are low)
Beams:	
ACI 318 (ACI 2002)	Beam clear span ≥ four times effective depth of beam; Beam width-to-depth ratio ≥0.3; Beam width ≥254 mm; ≥ width of supporting member (on plane perpendicular to beam axis) plus distances on each side of not greater than $\frac{3}{4}$ of overall beam depth
Eurocode 8 (CEN 2004)	Beam width ≤ $(b_c + h_w)$; ≤ $2b_c$; Centroidal axes of beam and column must not be more than $(b_c/4)$ apart; b_c = largest c/s dimension of column perpendicular to beam; h_w = depth of beam

Note: c/s = cross-section.

benefit from the improvement that column compression has on the bond of beam reinforcement passing through the joint region; the restrictions effectively prohibit the use of flat slab systems as ductile frames, since they perform poorly under earthquake loading. Depth-to-width ratios within individual elements are restricted to prevent buckling instability. Low beam span-to-depth ratios are likely to result in members governed by shear rather than flexure. This will restrict their ductility unless special measures are taken, such as provision of diagonal steel.

Subsection 5.4.3 in Chapter 5 set out some of the factors influencing overall frame geometry. Preliminary design usually then follows on an iterative basis using an equivalent static analysis, to establish that the chosen sections can be reinforced for the strength required, and that the stiffness is adequate. Often, stiffness rather than strength may govern the design of tall buildings. The process needs to be iterative because changing the stiffness of the structure changes its period of vibration, and hence the seismic loads it attracts.

Capacity design considerations are also important, even at preliminary planning stage, to ensure that favourable yielding mechanisms apply. In particular, a 'strong column–weak beam' frame should be assured to prevent soft or weak storey mechanisms forming during an earthquake, and the shear strength of an element should in most cases exceed that required to develop its flexural strength.

To satisfy the 'strong column–weak beam' condition, Eurocode 8 requires the sum of design column flexural strengths to exceed 130% of the sum of beam flexural strength framing into a joint, except on the top storey of a frame where the requirement is waived, while the corresponding strength ratio in IBC is 120%. In calculating the column flexural strength, due allowance must be made for the most unfavourable axial load that may be present. In calculating the beam flexural strength, the contribution of adjacent floor slabs should also be considered; this contribution may be considerable (Fenwick *et al.* 2005). Eurocode 8 alternatively allows a non-linear static (pushover) analysis to check that the hierarchy of beam and column strengths is satisfactory, and that weak storey mechanisms or other brittle failure modes are avoided. More rigorous and complex procedures are given in the commentary to the New Zealand concrete code NZS 3101 (NZS 1995).

To some extent, these capacity design considerations for relative flexural strength and shear strength can be satisfied by adjusting the amount of reinforcement. However, preliminary design needs to ensure that the section sizes chosen are sufficient to accommodate the required reinforcement without undue congestion.

8.7.3 Detailing of beams and columns

In order to ensure satisfactory seismic performance, careful detailing of reinforcing bars is essential, and codes of practice provide extensive guidance. Figures 8.13 and 8.14 show typical details for beams and columns respectively while Table 8.4 provides a summary of ACI and Eurocode requirements for ductile members.

8.7.4 Beam–column joints

The joint region between beams and columns in a moment-resisting frame is a highly stressed region, in which the shear stresses are many times greater than

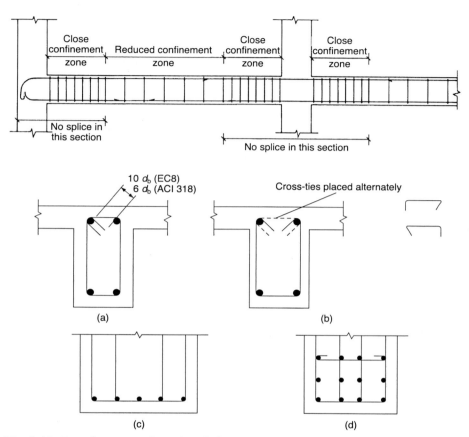

Fig. 8.13 Detailing notes for a ductile beam: (a) closed hoop; (b) stirrups with ties; (c) multi-leg hoops for wide beam; and (d) multiple layers of flexural steel

those in a frame subjected solely to gravity loads (Fig. 8.15). These high shear forces lead to high concrete diagonal compressive forces, which require good confinement of the joint region to be sustainable, and the need for horizontal and vertical shear steel to transmit the diagonal tension. They also imply a high rate of change in bending moment and hence lead to rapid changes in the tension forces in the flexural steel. The bond stresses between flexural steel and concrete in the joint zone are therefore also exceptionally high; bars passing through the joint are expected to be in full compressive yield on one side of the joint and in full tension yield on the other. This leads to the need to restrict the diameters of such bars (since bond resistance per unit length decreases with increasing bar diameter) and the need to provide good confinement to the bars, to sustain the high bond stresses which develop. Joints at the ends of beams also need special care, because the anchorage length for the beam steel on one side of the joint is restricted (Fig. 8.16). Note that none of the steel arrangements are particularly easy to fix; for example, the beam stubs shown in the lower left of the figure may require the main bars to be introduced from the outer edge.

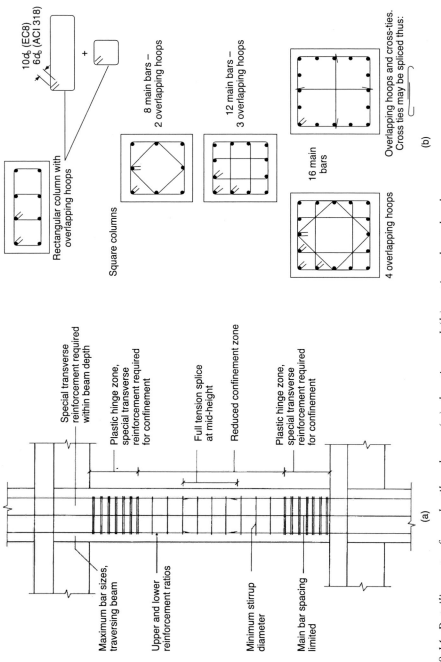

Fig. 8.14 Detailing notes for a ductile column: (a) elevation; and (b) sections through column

Table 8.4 Code detailing requirements for ductile beams and columns

Beams: main reinforcement:

ACI 318 (ACI 2002)	Ratio of main reinforcement $\geq 1.38/f_y$ where f_y is the yield strength of main steel Ratio of main reinforcement $\leq 2.5\%$ (Ratio of reinforcement is normalised to $b_w d$ where b_w is the beam width and d is the effective depth) At least two bars must be provided at top and bottom of the beam throughout its length Positive bending strength throughout beam $\geq 50\%$ of negative bending strength at joints Positive and negative strength everywhere in beam $\geq 25\%$ maximum bending strength at joints No lap joints are allowed: (*a*) within beam column joints (*b*) within 2 times member depth from joint face (*c*) within anticipated plastic hinge zones Laps must be confined by hoops or stirrups spaced at not more than $d/4$, or 102 mm, if less, where d is the effective depth
Eurocode 8 (CEN 2004) DCM structures	Ratio of main tension reinforcement $\geq 0.5 f_{ctm}/f_{yk}$ where f_{ctm} is the mean tensile strength of concrete and f_{yk} is the characteristic yield strength of steel Ratio of main tension reinforcement $$\leq \rho_{max} = \rho' + \frac{0.0018}{\mu_\varphi \varepsilon_{sy,d}} \times \frac{f_{cd}}{f_{yd}}$$ where ρ' is the ratio of compression steel; f_{cd}, f_{yd} is the design strength of concrete and steel respectively (i.e. characteristic strength divided by the appropriate partial material factor, γ_m); μ_φ is the curvature ductility ratio (typically around 7 for DCM structures and 11 for DCH structures); and $\varepsilon_{sy,d}$ is the design strain of reinforcement at yield (typically 0.22%) (Ratio of reinforcement is normalised to bd, where b is the width of the compression flange of the beam and d is the effective depth) At least half the area of tension steel is provided in compression zones, in addition to any design compression steel No lap joints are allowed: (*a*) within beam column joints (*b*) within anticipated plastic hinge zones Laps must be confined by hoops or stirrups spaced at not more than $h/4$, or 100 mm, if less, where h is the minimum cross-sectional dimension

Beams: transverse reinforcement:

ACI 318 (ACI 2002)	In the special confinement zone, spacing of hoops must not exceed:
	  $d/4$
	  8 times diameter of smallest longitudinal bar
	  24 times diameter of hoop bars
	  305 mm
	Outside this zone, spacing may be relaxed to $d/2$
	Hoops may consist of closed hoops, or stirrups and cross-ties (see Fig. 8.13)
	Hoops and stirrups must be terminated with a 135° hook, which extends 6 hoop bar diameters (or 76 mm if less) into the confined core of the beam

Table 8.4 Continued

Beams: transverse reinforcement (*continued:*)

Eurocode 8 (CEN 2004) DCM structures	In the special confinement zone, spacing of hoops must not exceed: *Total* beam depth/4 8 times diameter of smallest longitudinal bar 24 times diameter of hoop bars 225 mm Hoops may consist of closed hoops, or stirrups and cross-ties (see Fig. 8.13) Hoops and stirrups must be terminated with a 135° hook, which extends 10 hoop bar diameters into the confined core of the beam

Columns: main reinforcement:

ACI 318 (ACI 2002)	Ratio of main reinforcement $\geq 1\%$ $\leq 6\%$ Lap splices shall occur only in the centre half of the column
Eurocode 8 (CEN 2004)	Ratio of main reinforcement $\geq 1\%$ $\leq 4\%$ At least one intermediate bar shall be provided between column corners Symmetrical sections shall be reinforced symmetrically

Columns: transverse reinforcement:

ACI 318 (ACI 2002)	Spacing of column hoops or spirals in special confinement zone $\leq \frac{1}{4}$ minimum dimension of column ≤ 6 times diameter of smallest longitudinal bar $102\,\text{mm} \leq 102 + (356 - h_x)/3 \leq 152\,\text{mm}$ where h_x is the horizontal spacing between hoop or cross-tie legs and $h_x \leq 356\,\text{mm}$ Outside special confinement zone, spacing can be relaxed to 6 times diameter of smallest longitudinal bar or 152 mm if less Height of special confinement zone $\leq$ depth of column at joint face $\leq$ section over which yielding is expected $\leq$ 1/6 of clear span of column $\leq$ 457 mm
Eurocode 8 (CEN 2004)	Minimum diameter of hoops, ties or spirals is $0.4 d_{bL}\sqrt{f_{ydL}/f_{ydw}}$ where d_{bL} is the diameter of the main column bars and f_{ydL}/f_{ydw} is the ratio of yield strength in the main bars to that in the hoops Spacing of column hoops or spirals in special confinement zone $\leq$ 1/3 minimum dimension of confined core of column (to centre line of hoops or spirals) $\leq$ 6 times diameter of smallest longitudinal bar $\leq$ 125 mm Distance between main bars engaged by hoops or cross ties $\leq$ 150 mm Outside special confinement zone, Eurocode 2 (i.e. non-seismic) rules apply Height of special confinement zone $\leq$ 1.5 times the largest c/s dimension of column $\leq$ section over which yielding is expected $\leq$ 1/6 of clear span of column $\leq$ 600 mm The entire column shall be treated as a special confinement zone where the ratio (clear column height)/(max column c/s dimension) is less than 3

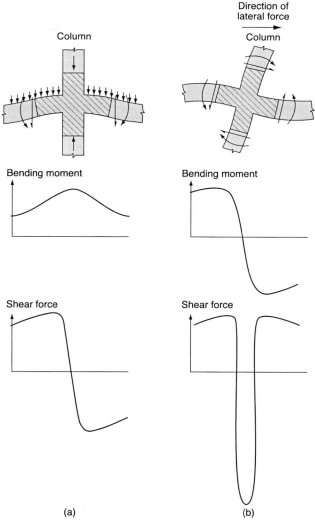

Fig. 8.15   Shear in beam column joints: (a) gravity frame; and (b) sway frame compared

A full and clear discussion of the complex transmission of forces within beam–column joints is provided by Paulay (1994).

In the rules for the design of beam–column joints, there is a clear distinction between the more rigorous approach of the New Zealand standard NZS 3101 (NZS 1995) on the one hand and US practice, represented by ACI 318 (ACI 2002), on the other. Eurocode 8 (CEN 2004) stands somewhere in between, with rigorous rules for DCH (high-ductility) structures, and much simpler ones for DCM (medium-ductility) structures. Debate on this issue is not entirely resolved.

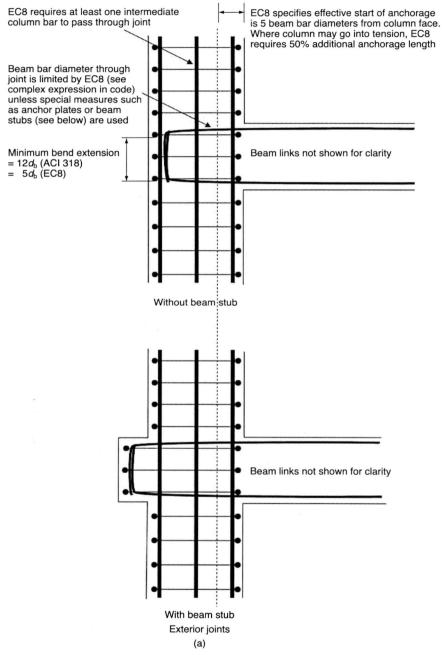

EC8 requires at least one intermediate column bar to pass through joint

EC8 specifies effective start of anchorage is 5 beam bar diameters from column face. Where column may go into tension, EC8 requires 50% additional anchorage length

Beam bar diameter through joint is limited by EC8 (see complex expression in code) unless special measures such as anchor plates or beam stubs (see below) are used

Minimum bend extension
= $12d_b$ (ACI 318)
= $5d_b$ (EC8)

Beam links not shown for clarity

Without beam stub

Beam links not shown for clarity

With beam stub
Exterior joints
(a)

Fig. 8.16 Anchorage of flexural steel in beam–column joints

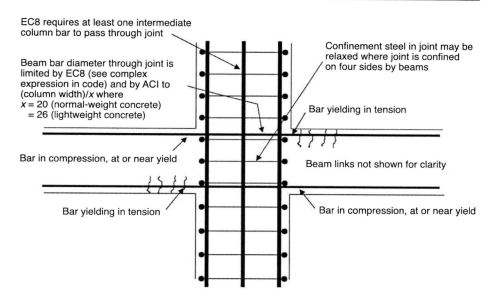

EC8 requires at least one intermediate column bar to pass through joint

Confinement steel in joint may be relaxed where joint is confined on four sides by beams

Beam bar diameter through joint is limited by EC8 (see complex expression in code) and by ACI to (column width)/x where
x = 20 (normal-weight concrete)
 = 26 (lightweight concrete)

Bar yielding in tension

Bar in compression, at or near yield

Beam links not shown for clarity

Bar yielding in tension

Bar in compression, at or near yield

Beam bars continuous through joint

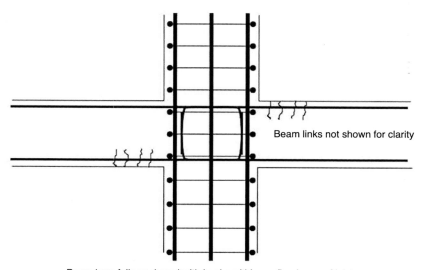

Beam links not shown for clarity

Beam bars fully anchored with hooks within confined area of joint
Interior joints
(b)

Fig. 8.16 Continued

It is generally accepted that either approach should provide sufficient strength in the joint to allow plastic hinges to form in the beams, allowing the frame to develop its potential ductility. It is however argued by proponents of the more rigorous methods that the US approach may in some cases lead to significant stiffness degradation in the joint under severe cyclic loading, leading to increased storey drift and associated damage.

8.7.5   Frames of 'low' or 'ordinary' ductility

The normal strength and detailing requirements of codes for structures designed to withstand wind and gravity loads in themselves provide some basic level of robustness and hence seismic resistance and ductility, which may be sufficient for areas of low seismicity. Such frames without seismic detailing or capacity design are recognised by Eurocode 8 as having 'low' (DCL) ductility and by ACI 318 as having 'ordinary' ductility. In Eurocode 8, they are designed for seismic forces calculated using the low behaviour factor q of 1.5 (compared with up to 6.75 for DCH frames) but are only recommended for areas of low seismicity. Similarly, ACI 318 specifies increased seismic forces for 'ordinary' frames, but only allows them in low-seismicity areas.

8.7.6   Frames not proportioned to resist lateral loads

Some frames may be designed to resist only the gravity loads in a building. This would be the case for the internal structure where lateral resistance is provided by a separate perimeter frame (Fig. 5.1(b)) or where shear walls take all the seismic loads. The design and detailing of gravity-only frames can clearly be less demanding than for the moment-resisting frames discussed so far; their only requirement is to maintain their load-carrying capacity under the maximum deflections imposed on them during an earthquake. The stiffer the seismic load-resisting system, the lower the deflection demanded, which is one reason why shear walls offer such good seismic protection.

Where the gravity-only members (and in particular the columns) are not expected to exceed their design flexural and shear strength under the imposed seismic deflections, less stringent measures are needed. For this case, ACI 318 places some restrictions on spacing of confinement steel in columns, which increase with the level of axial load. In cases where the compressive stress due to the axial load exceeds $0.1f'_c$, a capacity design for shear is needed, whereby the shear strength must be greater than that needed to develop the flexural strength at the ends of the column.

For the more critical case where the flexural strength of the gravity-only column is exceeded under the design seismic deflections, ACI 318 requires more stringent measures, which amount to full confinement steel as for ductile moment-resisting frames where the axial compressive stress exceeds $0.1f'_c$. The rules in Eurocode 8 are less stringent; no additional detailing beyond the non-seismic rules of Eurocode 2 is needed if the design deflections do not cause the gravity-only columns to yield but full ductile detailing is (apparently) required if yielding does occur. Some caution is needed in applying these rather relaxed Eurocode 8 rules in the case of no yield.

8.7.7 Precast concrete frames

The major potential source of weakness in precast frames lies in their connections. If capacity design procedures are carried out to ensure that yielding does not occur here, the rest of the structure is to be designed to rules for cast-in-situ structures. Alternatively, the connections can be specially designed to yield and provide energy dissipation under extreme seismic loading (although this is more likely to be practical in precast wall systems). Both approaches are recognised by Eurocode 8, which provides detailed design rules. Design guidance is also provided by a New Zealand set of guidelines (NZ Concrete Society 1999). Precast concrete frames are permitted in ACI 318 but detailed design rules are not given.

8.7.8 Moment-resisting frames with masonry infill panels

Eurocode 8 recognises three situations where external frames have been infilled with masonry walls. In the first, the walls are separated from the frames, so that there is no structural interaction between them. In practice, this is quite difficult to achieve, because the walls need lateral restraint to prevent them from falling out of the frame under strong wind or earthquake loading.

In the second situation, the walls are built up to the column members but are not connected to them. It must then be ensured that no weak storeys or short columns (Figs 1.5 and 1.17) are formed, and that the concrete frame can take the additional forces to which the infill panel may subject it. Eurocode 8 provides rules for this check.

The third situation is where the walls are built first, and the beam–column frame is cast directly against the masonry. The system is then treated as a 'confined masonry' structure, where all the seismic forces are considered to be resisted by the masonry, but enhanced resistance may be assumed, provided the reinforced concrete elements conform to certain minimum requirements (see Chapter 10).

8.8 Shear walls

8.8.1 Preliminary sizing

Shear walls must be thick enough to prevent buckling instability occurring under extreme seismic loading, and must also usually be able to accommodate two horizontal and two vertical layers of reinforcement. Eurocode 8 (CEN 2004) requires a minimum web thickness of 150 mm, or $(h_s/20)$ where h_s is the clear height of the wall.

In the lower part of the wall, where a plastic hinge would be expected to form, there are particularly great demands on the outer edges of the wall, which are known as boundary elements. These need to accommodate the flexural tension steel, and also confinement steel to sustain the concrete compressive strains. Eurocode 8 requires a minimum thickness of 200 mm and between $(h_s/15)$ and $(h_s/10)$ in these boundary elements, depending on their length.

8.8.2 Flexural and shear strength of slender shear walls

As with the design of beam–column joints, there is a clear distinction between simpler and more empirical US practice, and the more complex and rigorous

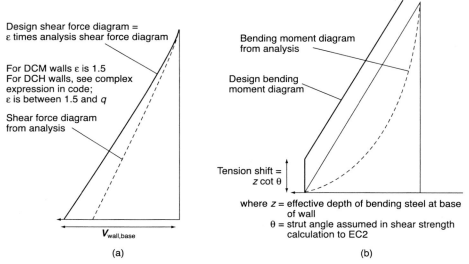

Design shear force diagram =
ε times analysis shear force diagram

For DCM walls ε is 1.5
For DCH walls, see complex
expression in code;
ε is between 1.5 and q

Shear force diagram
from analysis

$V_{wall,base}$

(a)

Bending moment diagram
from analysis

Design bending
moment diagram

Tension shift =
$z \cot \theta$

where z = effective depth of bending steel at base
of wall
θ = strut angle assumed in shear strength
calculation to EC2

(b)

Fig. 8.17   (a) Shear force and (b) bending and distributions from Eurocode 8 for design of isolated shear walls

procedures of Eurocode 8 (particularly for DCH structures) and the New Zealand concrete code.

US practice takes design bending and shear forces directly from analysis, without any capacity design considerations. Flexural strength is determined exactly as for beams or columns, while shear strength is based on a simple formula depending on the amount of web reinforcement, the concrete strength and the aspect ratio of the wall. Often, a shear wall is designed as contained within a beam–column frame; the frame is sized to support the tributary gravity loads without any support from the wall. The structure can then be treated in IBC (ICC 2003) as having a separate frame, rather than as being a 'bearing wall system', and attracts a more favourable structural or R factor.

There are two important differences from this US approach in Eurocode 8. First, both bending moments and shear forces are not based directly on the seismic forces obtained from the analysis, but on a capacity design approach. This is intended to ensure that the flexural hinges form only in the lower part of the wall and also that the shear strength everywhere exceeds the value needed to develop the wall's flexural strength. Figures 8.17 and 8.18 show how the bending moment and shear force distributions obtained from analysis relate to the distributions used for checking design strength. The design shear force diagram exceeds the distribution obtained from analysis, first because of the (uncertain) influence of higher mode effects (which are relatively much more important, compared with bending moments). Second, it also allows for the more classic capacity design consideration, that the flexural strength provided at the base of the wall will in general exceed the analysis value, hence allowing correspondingly large shear forces to develop. The 'tension shift' in the bending moment diagram arises purely from static considerations; as shown in Fig. 8.19, the force in the flexural steel at the *top* of

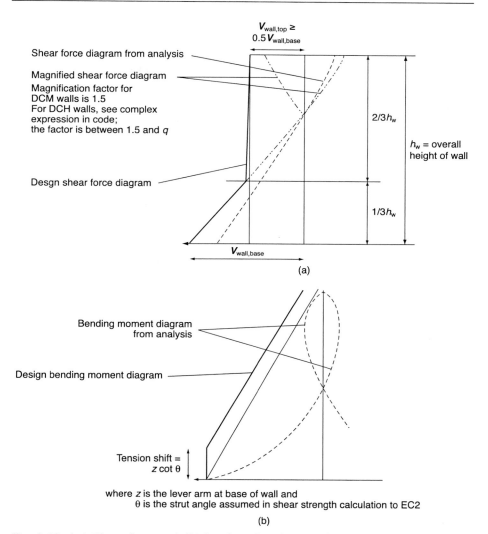

$V_{wall,top} \geq$
$0.5 V_{wall,base}$

Shear force diagram from analysis

Magnified shear force diagram

Magnification factor for
DCM walls is 1.5
For DCH walls, see complex
expression in code;
the factor is between 1.5 and q

$2/3 h_w$

h_w = overall
height of wall

Desgn shear force diagram

$1/3 h_w$

$V_{wall,base}$

(a)

Bending moment diagram
from analysis

Design bending moment diagram

Tension shift =
$z \cot \theta$

where z is the lever arm at base of wall and
θ is the strut angle assumed in shear strength calculation to EC2

(b)

Fig. 8.18 (a) Shear force and (b) bending distributions from Eurocode 8 for design of shear walls forming part of dual systems

a diagonal crack is more closely related to the bending moment at the *bottom* of the same crack, giving rise to the tension shift in Figs 8.17 and 8.18. The straightening of the bending moment diagram allows (roughly) for higher mode effects. In the New Zealand code (but not the Eurocode), the bending strength in upper sections of the wall must be further increased to allow for the flexural strength actually provided at the base, which will in general exceed the value from the analysis.

Different requirements apply to frame-wall or dual systems (where shear walls combine with frames to provide lateral resistance – see Fig. 8.24) because of the complex interaction between frames and walls. A more detailed discussion of these factors is given by Penelis and Kappos (1997).

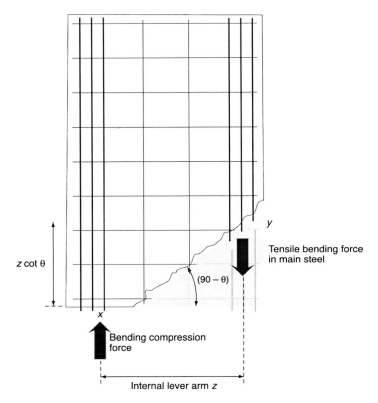

z cot θ

(90 − θ)

y

Tensile bending force
in main steel

x

Bending compression
force

Internal lever arm z

The figure shows the free body diagram above a main shear crack at the base of
a shear wall. Neglecting the contribution of the web steel crossing the crack to
the bending moment at the base and also of the concrete interfaces in the crack
(both of which are relatively small), it can be seen that the bending force in the
main steel at the top of the crack at level y is most closely related to the bending
moment at the bottom of the wall level x. The difference between levels x and y is
effectively the 'tension shift' shown in Figs 8.17 and 8.18.

Fig. 8.19   Tension shift mechanism in shear walls

Thus, there is a significant difference between US and Eurocode practice in
obtaining design moments and shears in slender shear walls, but there is also a
second difference which applies to the assessment of shear strength. In both ACI
318 (ACI 2002) and Eurocode 8, flexural strength of shear walls is assessed as
for beams and columns, on a 'plain sections remain plain' basis. However,
where the shear strength of walls in ACI 318 is determined from a simple one-
line formula based on the steel and concrete strengths and areas, and the wall's
aspect ratio, for DCH structures in Eurocode 8 separate checks are specified for
all the failure modes discussed in subsection 8.2.7. Once again, the procedure for
DCM structures is much simpler.

There is no clear evidence from past earthquakes that US practice in shear wall
design is deficient. However, Eurocode 8 and New Zealand requirements are based
on extensive experimental and analytical work, and the prudent (or cautious)

designer may find comfort in checking important or unusual shear walls to these more rational procedures, even when the more empirical rules are specified.

8.8.3   Boundary elements

Boundary elements are needed to sustain the highly stressed edges of shear walls in plastic hinge regions. Where the concrete strains exceed around 0.35% under the design seismic loads, confinement steel is required, similar to that specified for the critical regions of columns and taking the form of closed horizontal loops or horizontal cross-ties. Both US and Eurocode practice are clearly based on this principle, but once again the execution is rather different. For DCH structures, Eurocode 8 requires a 'first principles' approach in which the concrete strains are determined from a 'plain sections remain plane analysis' for the local ductility demand determined from the behaviour factor q used in design, and the ratio of flexural strength provided to that obtained from analysis. Appropriate confinement steel is then specified in the region of the wall where the concrete strain exceeds 0.35%. The rules for DCM are somewhat simpler. ACI 318 allows this approach but also gives a single closed-form equation to determine whether or not confinement steel is needed. Where the edge of the wall runs into an orthogonal wall to form a T or L section, the concrete strains are reduced, and the 'first principles' approach allows for the favourable effect of this.

8.8.4   Squat shear walls

As discussed in subsection 8.2.9, squat shear walls with a height-to-length ratio of less than around 2 behave in a rather different way to slender walls. ACI 318 recognises the increased shear strength available, and the need to provide diagonal reinforcement to resist high levels of shear. Eurocode 8 provides more detailed procedures for calculating shear strength, also requiring diagonal steel where the shear stress is high. The tension shift and higher mode effects are less important in squat shear walls than in slender ones, so the only capacity design requirement is to increase the shears from analysis by the ratio of actual base moment to base moment from analysis.

8.8.5   Openings in shear walls

Functional reasons often necessitate openings in shear walls. These will affect the flow of forces through the walls, and hence their strength. If they are large enough and placed in critical locations, they can form potential failure triggers (Fig. 8.20).

In the case of regular vertical arrays of openings, this can be turned to the designer's advantage by the formation of coupled shear walls (Fig. 5.8 in Chapter 5). The most detailed rules for the design of coupled shear walls, including capacity design requirements, are provided in the New Zealand code NZS 3101 (NZS 1995), although coupled shear walls are referred to in both Eurocode 8 and ACI 318. The coupling beams will usually require diagonal reinforcement, and the resulting eight layers of reinforcement results in a minimum practical thickness of 300 mm.

Code advice for design of openings in other cases is limited. In a ductile shear wall, the objective is to ensure that the weakening associated with introduction

Fig. 8.20   Damage around opening in a shear wall, 1985 Chile earthquake

of the shear wall does not reduce shear and bending strength locally below the level corresponding to plastic hinge formation at the base of the wall. Local flexural strength can be determined on a 'plane sections remain plane' assumption, while ACI 318 can be used to assess the overall shear strength of the wall as the sum of contributions from individual 'wall piers' (Fig. 8.21). Alternatively, a 'strut-and-tie' approach may be used, as described by Paulay and Priestley (1992). It is unlikely that significant openings can be accommodated near the outside edges of shear walls in the plastic hinge region without reducing the available ductility.

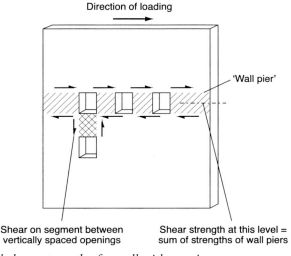

Fig. 8.21   Local shear strength of a wall with openings

8.8.6   Large panel precast buildings

Assembly of medium- to high-rise buildings from precast storey-height concrete wall panels offers the advantage of rapid site construction and casting of concrete under well-controlled factory conditions. It was extensively used in Western Europe in the 1960s and 1970s and more recently in Eastern Europe. A rather different system, called 'tilt-up' construction is found in the USA and New Zealand, where the walls to low-rise buildings are cast horizontal on site, and are then tilted up once cured to form the sides of the building.

The engineering problem with such construction lies in making adequate connections between the precast units; the potential for loss of a panel leading to progressive collapse has been recognised at least since the collapse of the Ronan Point building in England after a gas explosion (Griffiths *et al.* 1968). However, the record of large panel buildings in the 1978 Bucharest earthquake (Bouwkamp 1985) and 1988 Armenian earthquake (Wyllie and Filson 1989) is good, despite some very poor construction. Tilt-up buildings in California have proved more vulnerable, often failing at the connection between roof and wall (Fig. 8.22). Also, the wall thickness can be quite small; buckling and out-of-plane bending failures may govern.

Four main approaches may be used in the design of precast wall systems, as described by Mattock (1981).

(1) *Ductile cantilever shear walls*, which dissipate energy by plastic hinge formation at the wall bases, exactly as for a monolithic structure. In this case, horizontal joints must be strong enough to transmit moments and

Fig. 8.22 Failure of tilt-up warehouse building, Loma Prieta earthquake, California, 1989

shears, determined on a capacity design basis, and vertical joints must be capable of transmitting a similar level of shear.

(2) *Energy-dissipating connections*, where some of the connections are treated as ductile or semi-ductile fuses, limiting the forces on the other connections and the rest of the structure. Generally, the ductile connections will be placed in the vertical joints, since it is unlikely that sufficient distribution of ductility demand could be achieved between ductile connections on horizontal joints.

(3) *Elastically responding structures*, where sufficient strength is provided to prevent inelastic response in either structure or connections under the most severe anticipated ground motions. The use of base isolation may be an alternative means of limiting structural response to within elastic limits.

(4) *Energy-dissipating panels*, where special panels are introduced which absorb energy under severe seismic loading. This type of system has been developed and used in Japan.

Eurocode 8 provides advice on the first three of these approaches. Hamburger *et al.* (1988) give guidance on the design of tilt-up buildings.

8.9   Concrete floor and roof diaphragms

8.9.1   Structural functions of diaphragms

The floors and roof of a building, in addition to resisting gravity loads, are also generally designed to act as diaphragms. In this respect, they are required both to distribute seismic forces to the main elements of horizontal resistance, such as frames and shear walls, and also to tie the structure together so that it acts as a single entity during an earthquake. The robustness and redundancy of a structure is highly dependent on the performance of the diaphragms; Wyllie and Filson (1989) have suggested that the inadequacy of the floor diaphragms played an important part in the catastrophic performance of precast concrete buildings in the 1988 Armenian earthquake.

The seismic forces in a diaphragm may arise from two distinct causes, namely 'local' or 'transfer', and it is important to distinguish between the two, as follows.

(a) *Local forces* are those arising from the inertial loads at the level of the diaphragm, which need to be taken back to the main horizontal load-resisting structure.

(b) Diaphragms may also be required to resist *transfer forces* which arise where there is an offset in the horizontal load-resisting structure, for example at the transition level between a tower and podium (Fig. 8.23). Transfer forces may also arise in the case of a building where the lateral resistance is provided by both shear walls and frames. The transfer forces in the diaphragms arise from tying the walls and frames to describe the same vertical deflected shape, which separately would be different (Fig. 8.24).

8.9.2   Preliminary sizing of diaphragms

In-situ diaphragms are unlikely to be governed in size by seismic forces. However, in precast floor construction, seismic forces are usually transferred back to the

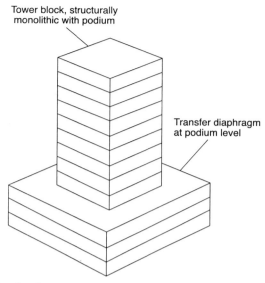

Fig. 8.23   Transfer diaphragm in a tower and podium building

lateral load-resisting structure solely through the in-situ topping on the precast units, which may therefore be highly stressed. ACI 318 requires the topping to be at least 50 mm thick, with 75 mm thickness required at the edges in some cases. Eurocode 8 specifies a minimum thickness of 70 mm and minimum reinforcement in two directions. Precast floors without an in-situ topping are not generally recommended in seismic areas.

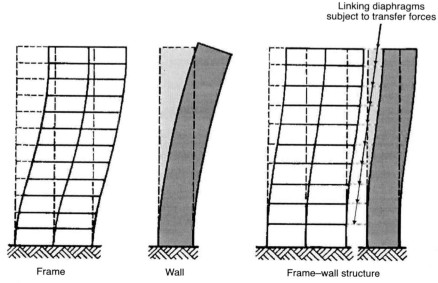

Fig. 8.24   Transfer diaphragms in a frame wall or dual-system building

8.9.3   Analysis for final member sizing

(a) Capacity design considerations

In a ductile structure, diaphragms will almost always be required to remain elastic, so that they can sustain their function of transferring forces to the main lateral-resisting structure, and tying the building together. Diaphragms should in principle therefore have the strength to sustain the maximum forces that may be induced in them by the chosen yielding mechanism within the rest of the structure.

Eurocode 8 deals with this rather simply by specifying that diaphragms should be designed for 1.3 times the shear forces obtained directly from the analysis. IBC has no direct capacity design requirement, although as noted below its requirements in other respects are rather more stringent than those of Eurocode 8.

(b) Diaphragm flexibility

Usually, the seismic analysis of buildings is carried out on the assumption that deflections in the diaphragms are so small compared with those in the main lateral load-resisting structure that the diaphragms can be treated as rigid. In most cases, this is quite satisfactory, because usually diaphragm flexibility affects neither overall structural stiffness (and hence natural period) nor the distribution of forces within a structure. Moreover, during a major earthquake, in ductile structures where the diaphragms are designed to remain essentially elastic, the superstructure deflections are likely to include large plastic deformations, increasing the disparity still further. However, diaphragm flexibility can be important in two cases, as follows.

(1) In structures with 'transfer' diaphragms, where the flexibility can significantly affect the distribution of load between lateral resisting systems.
(2) When considering serviceability limit states in buildings with relatively flexible diaphragms, because the lateral load-resisting system would be expected to remain elastic, and so remain with a comparable stiffness to the diaphragms.

(c) Local and transfer forces

Local diaphragm forces are likely to be appreciably greater than the code-prescribed equivalent static forces at each level, because the latter reflect the change in peak shear force at each level, whereas higher-mode effects may give rise to accelerations causing greater local forces. The special procedures for assessing diaphragm forces given in IBC recognise this effect, though it is not considered in Eurocode 8.

Transfer forces should in principle be based directly on capacity design considerations, based on the as-built strength of the potential yielding zones of the structure, with due allowance for strain hardening in steel. In practice, neither IBC nor Eurocode 8 requires this, although the New Zealand concrete code NZS 3101 (NZS 1995) does.

IBC recognises the difference between local and transfer forces and requires them to be added together for the purposes of design. This is likely to be conservative, because they are arise from different modes of vibration within the structure, and an SRSS (square root of the sum of the squares) combination is more likely to be appropriate. Eurocode 8 provides no advice in this respect.

8.9.4 Strength of diaphragms

ACI 318 (ACI 2002) specifies that the shear strength of diaphragms should be assessed as for shear walls. When assessing flexural strength, the bending forces are assessed as being concentrated in boundary elements at the edges of the diaphragm.

Eurocode 8 allows diaphragm strength to be assessed as a deep beam or by 'strut-and-tie' methods, whereby the tie forces are taken in the reinforcement and the concrete provides compression struts. Strut-and-tie methods are particularly useful in the presence of openings in diaphragms; they are further described by Schlarch *et al.* (1987) and Collins and Mitchell (1987).

8.10 Unbonded prestressed construction

A recent development has been research into the use of precast concrete frames connected through the beams by prestressed, post-tensioned cables which are left unbonded around the joint zones. Under high seismic excitation these joints open up, greatly increasing the flexibility of the structure and detuning it from the earthquake forcing frequencies. Once the seismic excitation stops, the prestress in the cables causes the gaps to close and the structure to return to its original condition with little or no damage, although non-structural damage may be involved. The structure can be described as a non-linear but elastic system.

Such structures have lower levels of viscous and hysteretic damping than conventional frames. This means they experience greater peak deflections (typically 40%), which has implications for the design of cladding and other non-structural elements, but the increase is limited because of the detuning effects noted above. Supplemental dampers may be used to reduce deflections still further; alternatively some mild reinforcement can be added across the joints. The attractive feature of the system returning undamaged to its original position at the end of a severe earthquake makes it worth further consideration. It is described in Priestley (2003, ch. 4) and the 2006 revision to the New Zealand concrete code NZS 3101 is expected to contain an appendix giving design rules.

8.11 References

ACI (2002). ACI 318-02/318R-02. *Building code requirements for structural concrete and commentary*. American Concrete Institute, Detroit, MI.

ACI Committee 363 (1988). *State of the art report on high strength concrete. Manual of Concrete Practice*, Vol. 1, ACI 363R-84, American Concrete Institute, Detroit, MI.

Ahmad S. H. & Shah S. P. (1982). Stress–strain curves for concrete confined by spiral reinforcement. *ACI Journal*, Vol. 79, Issue 6, pp. 484–490.

ASCE (1998). ASCE 4-98. *American Society of Civil Engineers Standard: Seismic analysis for safety-related nuclear structures*. ASCE, Reston, VA.

BIAX (1992). A computer program for the analysis of reinforced concrete and reinforced masonry sections. Available from http://www.eerc.berkeley.edu/software_and_data/

Bouwkamp J. (1985). *Building Construction under Seismic Conditions in the Balkan Regions*. Vol. 2: Design and construction of prefabricated reinforced concrete buildings systems. UNDP/UNIDO project RER/79/1015, United Nations Industrial Development Organization, Vienna.

CEN (2004). EN1998-1: 2004. *Design of structures for earthquake resistance*. Part 1: General rules, seismic actions and rules for buildings. Part 2: Bridges. Part 3: Assessment and retrofitting of buildings. European Committee for Standardisation, Brussels.

Collins M. P. & Mitchell D. (1987). *Prestressed concrete basics*. Canadian Prestressed Concrete Institute, ch. 9, pp. 386–430.

Derecho A. T. & Kianoush M. R. (2001). Seismic design of reinforced concrete structures. In: Naeim F. (ed.) *The Seismic Design Handbook*. Kluwer Academic, Boston.

DRAIN-2DX (1993). Static and dynamic analysis of inelastic plane structures. Available from http://www.eerc.berkeley.edu/software_and_data/

FEMA (2000). FEMA 356. *Prestandard and commentary for the seismic rehabilitation of buildings*. Prepared by American Society of Civil Engineers for the Federal Emergency Management Agency (FEMA), Washington DC.

Fenwick R. (1994). The behaviour of reinforced concrete under cyclic loading. In: Booth E. (ed.) *Concrete Structures in Earthquake Regions*. Longman Scientific & Technical, Harlow, UK.

Fenwick R. & Davidson B. (1993). *Elongation in ductile seismic resistant reinforced concrete frames*. ACI Special Publication SP157. American Concrete Institute, Detroit, MI.

Fenwick R. & Megget L. (1993). Elongation and load deflection characteristics of reinforced concrete members containing plastic hinges. *Bulletin of the New Zealand Society for Earthquake Engineering*, Vol. 26, No. 1, pp. 28–41.

Fenwick R., Davidson B. & Lau D. (2005). Interaction between ductile RC perimeter frames and floor slabs containing precast units. *Proceedings of New Zealand Society for Earthquake Conference*, March, Wairakei, NZ.

Fenwick R., Dely R. & Davidson B. (1999). Ductility demand for uni-directional and reversing plastic hinges in ductile moment resisting frames. *Bulletin of the New Zealand Society for Earthquake Engineering*, Vol. 32, No. 1, pp. 1–12.

Griffiths H., Pugsley A. & Saunders O. (1968). *Report of the inquiry into the collapse of flats at Ronan Point, Canning Town*. Her Majesty's Stationery Office, London.

Hamburger R. O., McCormick D. L. & Hom S. (1988). Performance of tilt-up buildings in the 1987 Whittier Narrows earthquake. *Earthquake Spectra*, Vol. 4, Issue 2, pp. 219–254.

ICC (2003). IBC: 2003. *International Building Code*. International Code Council, Falls Church, VA.

Lau D., Fenwick R. & Davidson B. (2002). Seismic performance of r/c perimeter frames with slabs containing prestressed units. *Proceedings of New Zealand Society for Earthquake Conference*, Napier, Paper 7.2.

Mander J. B., Priestley M. J. N. & Park R. (1988). Theoretical stress–strain model for confined concrete. *ASCE Journal of Structural Engineering*, Vol. 114, Issue 8, pp. 1804–1826.

Mattock (1981). A survey of precast wall systems. In: Mayes R. (principal investigator) *ATC8: Proceedings of a Workshop on Design of Prefabricated Buildings for Earthquake Loads*. Applied Technology Council, Redwood City, CA.

Mo Y. (2003). Reinforced concrete structures. In: Chen W.-F. & Scawthorn C. (eds) *Earthquake Engineering Handbook*. CRC Press, Boca Raton, FA.

New Zealand Concrete Society and NZ Society for Earthquake Engineering (1999). *Guidance for the Use of Structural Precast Concrete Frames in Buildings*, 2nd edn. Centre for Advanced Engineering, University of Canterbury, NZ.

NZS (1995). NZS 3101: Part 1. *The design of concrete structures*. Standards New Zealand, Wellington, NZ. (Note: a revised edition is expected to be published in 2006.)

Otani S., Hiraishi H., Midorikawa M. & Teshigawara M. (2000). New seismic design provisions in Japan. *ACI Fall Conventions*, Toronto.

Park Y. J., Ang A. H.-S. & Wen Y. K. (1987). Damage-limiting aseismic design of buildings. *Earthquake Spectra*, Vol. 3, Issue 1, pp. 1–26.

Paulay T. (1994). *The Fourth Mallet-Milne Lecture: Simplicity and Confidence in Seismic Design.* Wiley, Chichester.

Paulay T. & Priestley M. J. N. (1992). *Seismic Design of Reinforced Concrete and Masonry Buildings.* Wiley, New York.

Paulay T., Priestley M. J. N. & Synge A. J. (1982). Ductility in earthquake-resisting squat shear walls. *ACI Journal*, Vol. 79, Issue 4, pp. 257–269.

Penelis G. G. & Kappos A. J. (1997). *Earthquake-resistant Concrete Structures.* E & FN Spon, London.

Priestley M. J. N. (2003). Myths and fallacies in earthquake engineering, revisited. *The Mallet Milne Lecture 2003.* IUSS Press, Pavia, Italy.

Priestley M. J. N., Seible F. & Calvi G. M. (1996). *Seismic Design and Retrofit of Bridges.* Wiley-Interscience, New York.

Richart F. E., Brandtzaeg A. & Brown R. L. (1928). *A study of the fracture of concrete under combined compressive stresses.* University of Illinois, Engineering Experimental Station, Bulletin No. 185.

Riva N. N. & Cohn M. Z. (1990). An engineering approach to non-linear analysis of inelastic response of r/c members. *ASCE Journal of Structural Engineering*, Vol. 118, Issue 8, pp. 2162–2186.

Schlarch J., Schafer K. & Jennewein M. (1987). Towards a consistent design of reinforced concrete. *Prestressed Concrete Journal*, Vol. 32, Issue 3, pp. 74–150.

Tjokrodimuljo K. (1985). *Behaviour of reinforced concrete under cyclic loading.* University of Auckland School of Engineering, Report No. 374.

Wyllie L. & Filson J. (eds) (1989). Armenian earthquake report. *Earthquake Spectra* special supplement.

9 Steelwork design

'Many practising engineers have believed for years, albeit incorrectly, that steel structures were immune to earthquake-induced damage as a consequence of the material's inherent ductile properties... However [recent earthquakes have] confirmed research findings that material ductility alone is not a guarantee of ductile structural behaviour.'

Michel Bruneau *et al*. In: *Ductile Design of Steel Structures.*
McGraw-Hill, 1998

This chapter covers the following topics.
- The lessons from earthquake damage
- The behaviour of steel members under cyclic loading
- Ductility in steelwork, and how to achieve it
- Material specification
- Special considerations for analysis
- Design and detailing of moment-resisting frames
- Design and detailing of concentric and eccentric braced frames

9.1 Introduction

Structural steel is in several ways an ideal material for earthquake resistance, possessing high material ductility and high strength-to-mass ratio. However, considerable care is needed in the design and detailing of steel structures in order to ensure that a ductile end-result is achieved under the conditions of extreme cyclic loading experienced during an earthquake. Special attention is needed in the design of connections (particularly welded connections) joining members intended to yield, and to compression struts intended to buckle during the design earthquake. Also, in general, steel structures tend to be more flexible than equivalent concrete structures and, unless controlled, the resulting larger displacements may lead to higher levels of damage to non-structural components and to more significant P–δ effects.

The behaviour of steel elements subjected to repeated plastic deformations is in many ways as complex as that of reinforced concrete elements. Seismic design provisions for steel in the 1970s and early 1980s tended to ignore these complexities, and the steel sections of seismic codes were very much shorter than those for concrete. Subsequent research, and failures during earthquakes, have resulted in much longer and more complex code provisions. The steel provisions of

Eurocode 8 Part 1 (CEN 2004) run to 24 pages, and the equivalent provisions of the AISC code (AISC 2002) from the USA are 32 pages long, even excluding appendices.

9.2 Lessons learned from earthquake damage

The collapse of the 21-storey Pinot Suarez building in the 1985 Mexican earthquake was probably the first example of failure in a modern welded steel frame building in an earthquake (EEFIT 1986). Failure appears to have initiated in axial loading in the welded steel plate box columns on the core, which was braced in a chevron or horizontal K pattern (Fig. 9.1). Despite this high-profile failure of a building dating from 1971, Yanev *et al.* could conclude in 1991, on the basis of studying steel building performance in 11 earthquakes between 1964 and 1990, that:

> 'Buildings of structural steel have performed excellently and better than any other type of substantial construction in protecting life safety and minimizing business interruption due to earthquake induced damage. The superior performance of steel buildings, as compared to buildings of other construction, is evident even in structures that have not been specifically designed for seismic resistance.'

However, subsequent events have forced this optimistic view to be tempered to some extent. Although there are undoubted intrinsic advantages in using steel in earthquake country, the widespread failures in steel buildings that occurred in the 1994 Northridge (California) and 1995 Kobe (Japan) earthquakes showed that steel was far from immune, seismically. It became evident that at least some

Fig. 9.1 Collapse of Pinot Suarez building, Mexico City, 1985

of the apparently good performance of steel buildings was due to the fact that very few had been severely tested in an earthquake before 1994. Elnashai *et al.* (1995a) concluded after the Kobe earthquake

> 'The behaviour of steel structures was on the whole disappointing. It confirmed the serious doubts raised in the Northridge earthquake regarding the adequacy of existing design guidance. It will take very considerable efforts to establish the causes of the observed damage patterns. It will take even longer to regain confidence in steel as the primary seismic resistance material, if at all.'

Many of the failures were associated with fractures initiating in the heat-affected zones of welds. Elnashai *et al.* (1995a) quote a report by the Architectural Institute of Japan, which specifies the main damage patterns observed in Kobe as follows

- (*a*) cracking at beam-to-column connections (very high incidence rate, up to 70%)
- (*b*) complete severance of members near the weld access hole
- (*c*) severe damage or failure of column bases (101 out of 218 buildings inspected!)
- (*d*) in a few cases, beam hinging was observed
- (*e*) fracture at the location of internal stiffeners
- (*f*) buckling of members and collapse at connections of tubular steel frames
- (*g*) fracture and overall buckling of slender bracing members.

In the 1994 Northridge earthquake in California, it appeared at first that the many tall steel buildings in the epicentral area had been undamaged. However, subsequent studies showed that around 200 modern steel moment frame buildings had suffered severe cracks in the welded-flange, bolted-web moment connection between beams and column (Fig. 9.2). The detail that proved vulnerable was

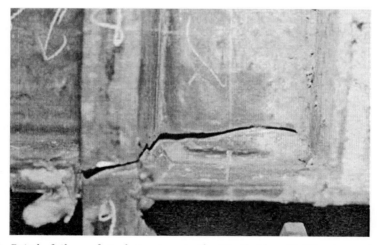

Fig. 9.2   Brittle failure of steel structure in the Northridge, California earthquake of 1994

widely used, and was recommended in the then current US seismic code. Engelhardt and Sabol (1996) concluded

> 'Based on the available evidence, no single factor has been isolated as the sole cause of the damage. Rather, it appears that a number of interrelated factors combined to cause the non-ductile failures of steel moment connections in the Northridge earthquake. Both welding related factors and a poor connection design appear to be the foremost among contributing factors. Problems with the welds included the use of low toughness filler metals combined with the presence of notches caused by welding defects and left-in-place backing bars. The basic connection design also contributed to the failures by generating excessively high stresses in the region of the beam flange groove welds. In addition to welding and design deficiencies, several other factors have been conjectured as playing some role in the failures. These include base metal factors, scale effects, composite floor slab effects and strain rate effects.'

Since the Northridge and Kobe earthquakes, there has been greater emphasis on improved weld filler material and weld detailing to remove stress concentrations, and also on designs that move the plastic hinging regions away from the immediate vicinity of welds, for example by reinforcement at the connections. Design solutions are discussed in greater detail later in this chapter.

9.3 The behaviour of steelwork members under cyclic loading

9.3.1 Introduction

Steel is a highly ductile material, and can achieve tensile and compressive strains of many times the yield strain (typically around 0.2% for high-yield steel) before fracturing. Two phenomena, however, may drastically limit the ductility that can be achieved in practical structures subjected to reversing load cycles well into the plastic range.

First, fractures may develop from points of stress concentration, particularly in the heat-affected zone next to welds, where the material ductility may have been reduced. This is the phenomenon of low-cycle fatigue, and led to many of the failures at Northridge and Kobe discussed above.

Second, buckling under compressive stress may reduce failure strength to well below that corresponding to yield. The effect is particularly marked for reversing loads because of the Bauschinger effect. This describes a fundamental property of steel, whereby its stiffness is reduced under loading in one direction if it has previously yielded due to loading in the opposite direction. Since buckling is governed by member stiffness, the buckling strength of a member is reduced if it has previously yielded in tension and each successive cycle reduces the stiffness, and so buckling strength, still further. This affects not only the overall buckling of a bracing member acting alternately as a tie and then a strut in successive loading cycles, but also local flange buckling of a plastic hinge region of a beam subject to reversing moments.

These two phenomena therefore have the potential both to reduce the ductility of a steel structure and to degrade its stiffness under successive load cycles.

9.3.2 Cyclic loading of struts

The behaviour of struts under reversing loads depends greatly on the slenderness ratio of the strut – that is, the ratio l/r_y of its effective length to radius of gyration along its weak axis.

A stocky strut is one in which yielding and local buckling dominate response. The maximum compressive load a stocky brace can sustain occurs when it yields in compression over its entire cross-section or suffers a local buckling failure. Typically, a stocky strut in Grade 50 steel has a slenderness ratio of 50 or less.

Figure 9.3 shows the response of a stocky strut to reversing loads. The compressive strength is somewhat less than the tensile strength, and reduces to some extent with successive cycles (by about a third in four cycles for the strut of Fig. 9.3), but there is very little reduction in stiffness.

A slender strut under compression is dominated by elastic buckling and will fail at a compressive load much less than its tensile strength. Typically, a slender strut has a slenderness ratio of 120 or more. Figure 9.4 shows the response of a slender strut to reversing loads. In contrast to the stocky strut, there is a large loss in compressive strength with successive cycles (over half in four cycles for the strut of Fig. 9.4), and a huge reduction in stiffness. The loss of both strength and stiffness on the compression cycle is very significant in the seismic response of braced frames, especially for V-braced and other frames which rely on both compression and tension members; this aspect is discussed in more detail in section 9.8.

The loss of strength and stiffness in slender struts is due both to the Bauschinger effect, described above, and to the strut becoming increasingly bent, even in the tension part of the loading cycle. This arises because of the following. Consider a slender strut subjected to cyclic loading sufficient to buckle the strut. In the compression cycle the strut will form a plastic hinge near midspan, and plastic hinge rotation will remain when the compressive force is removed (Fig. 9.5). Applying a tensile axial force can never completely remove this plastic rotation, even if the strut yields in tension. This is because a moment equal to the plastic hinge strength must be applied at the middle of the strut in order to reverse the

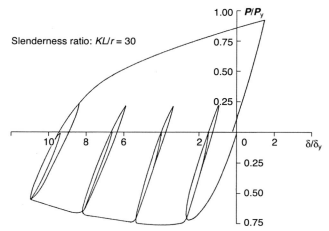

Fig. 9.3 Response of a stocky strut under cyclic loading (from Jain et al. 1978)

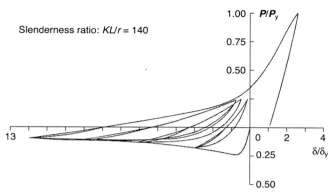

Slenderness ratio: $KL/r = 140$

Fig. 9.4 Response of a slender strut under cyclic loading (from Jain et al. 1978)

hinge's plastic rotation. The residual deformation cannot therefore become less than the plastic hinge strength divided by the tensile yield strength, for a pin-ended strut (Fig. 9.5). In many cases, the maximum tensile force on the strut will be much less than yield, and the residual plastic rotation of the hinge in the strut will increase with each loading cycle. This will successively reduce both compression stiffness and strength.

Struts of intermediate slenderness (slenderness ratio between about 50 and 120 for Grade 50 steel) respond in compression primarily by plastic buckling; that is, a plastic hinge forms soon after elastic buckling starts. The initial compressive strength is significantly less than the tensile strength, and reduces with successive cycles, as does the stiffness, but to a much lesser extent than is the case for

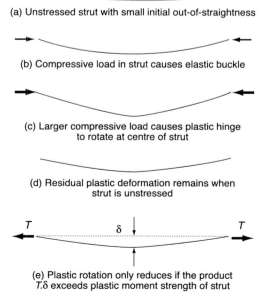

(a) Unstressed strut with small initial out-of-straightness

(b) Compressive load in strut causes elastic buckle

(c) Larger compressive load causes plastic hinge
to rotate at centre of strut

(d) Residual plastic deformation remains when
strut is unstressed

(e) Plastic rotation only reduces if the product
$T.\delta$ exceeds plastic moment strength of strut

Fig. 9.5 Residual deformations forming on cyclic loading of a slender strut

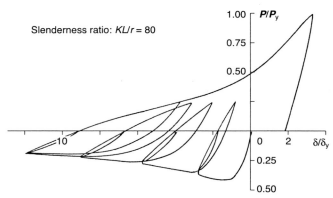

Fig. 9.6 Response of a strut with intermediate slenderness under cyclic loading (from Jain et al. 1978)

slender struts. Both the Bauschinger effect and residual deformations apply, but there is a smaller tendency for non-recoverable plastic residual deformations to develop, because the elastic stage of buckling produces relatively smaller lateral deformations (Fig. 9.6).

9.3.3 Cyclic loading in flexure
Assuming that premature weld failure does not occur, the cyclic behaviour under flexural loading is controlled either by local flange or web buckling or by lateral torsional buckling.

Onset of local flange buckling (Fig. 9.7) is governed by the local slenderness of the flange, which is one of the determinants of the 'compactness' of the section (the other determinant of compactness is the web slenderness). Eurocode 3 recognises four ranges of compactness

- (*a*) Class 1: Plastic cross-sections that can form a plastic hinge with significant rotation capacity
- (*b*) Class 2: Compact cross-sections that can develop their plastic capacity but with limited rotation capacity
- (*c*) Class 3: Semi-compact sections that can develop the yield moment but not the full plastic moment capacity of the cross-section
- (*d*) Class 4: Slender cross-sections that are unable to develop the yield moment due to early occurrence of local buckling.

For a rolled I-section beam, Eurocode 3 gives limiting $b_f/2t_f$ ratios of 10 for Class 1, and 11 for Class 2, for a yield stress $f_y = 235$ MPa, where b_f is the flange breadth and t_f the flange thickness. These figures drop to 8.1 and 8.9 for $f_y = 355$ MPa. Limits are also placed on the web slenderness for each compactness class, to control local web buckling. The AISC specifications in the USA have similar classifications.

Figure 9.8 shows the hysteretic response of a Class 1 section cycled to a ductility ratio of 7.2. The beam was laterally braced to prevent lateral torsional buckling. Local flange buckling was first observed during the second half cycle, but it can

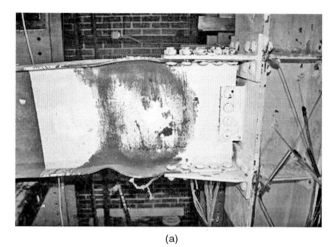

(a)

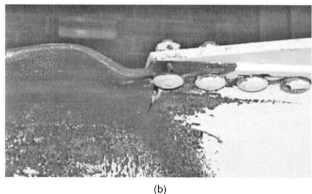

(b)

Fig. 9.7 (a) Local flange buckling and (b) fracture in a cyclically loaded flexural member

be seen that the beam survived to 12 cycles with a fairly gradual loss in strength and stiffness. The same mechanisms are responsible for this loss as was described above for cyclically loaded struts, namely the Bauschinger effect and increasing residual deformation, but they act on a local rather than member level. In dissipative structures, where plastic yielding is intended to occur to dissipate energy in the design earthquake, Eurocode 8 requires Class 1 or 2 sections, with Class 1 required to achieve a behaviour factor $q > 4$. These requirements apply to all members intended to achieve plasticity in the design earthquake ('dissipative members' in Eurocode 8 parlance), including flexural members in moment frames, compression struts in braced frames and ductile links in eccentrically braced frames.

Lateral torsional buckling is the phenomenon caused by overall instability of the beam's compression flange buckling between lateral restraint points. Onset is determined by a number of factors, principally the flange width-to-thickness ratio (b_f/t_f) and the slenderness ratio of the beam l/r_y (where l is the effective unrestrained length of the beam and r_y is the radius of gyration about the minor axis). Vann *et al.* (1973)

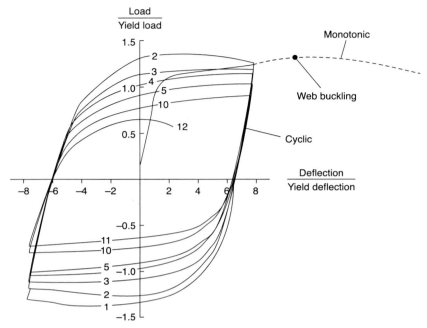

Fig. 9.8 *Compact steel member (b_f/2t_f = 7.8) loaded cyclically in flexure (after Vann et al. 1973)*

concluded that loss of stiffness is much more significant when lateral torsional buckling dominates response, rather than local web or flange buckling. They also found that deterioration of strength is only severe when local flange buckling is combined with either local web buckling or lateral torsional buckling. In addition to rules affecting compactness (and hence the b_f/t_f ratio), the US code AISC restricts maximum slenderness ratio in special ductility structures to $(17238/f_y)$. This gives $l/r_y \leq 73$ and 48 for $f_y = 235$ and 355 MPa respectively. Eurocode 8 places no further limitations on l/r_y for flexural members other than those given by Eurocode 3 for non-seismic situations; however, the effective length must be calculated assuming a plastic hinge forms at one end of the beam.

9.3.4 Cyclic loading of welds

Figure 9.2 shows the low-cycle fatigue failure of a weld. A number of factors tend to promote such failure, most importantly the following.

(a) The flexural strength of the connection of which the weld is a part is less than that of the beam it connects. Note that if the beam supports a floor slab, this will act as a flange, strengthening the flexural strength of the beam but most likely not the connection.

(b) The presence of stress raisers around the weld. These may include the effect of stopping and starting the bottom flange weld at the web, weld access holes and the effect of backing strips, if not removed and ground out.

(c) The weld metal is of low ductility.

(d) A welding procedure has been used which has led to brittleness in the heat-affected zone of the connected member.

Measures to avoid such factors applying are central to current code recommendations for the seismic design of welds, and are discussed in the subsection on connection design (9.10.6) below.

9.3.5   Scale effects

Two factors linked to scale effects have been suggested as contributing to the connection failures observed in the Northridge earthquake (Engelhardt and Sabol 1996). Both are connected with the Californian practice of using very large member sizes in the seismic-resisting elements, to minimise their number.

First, the use of very deep members in relation to the span means that, compared to more conventional sizes, relatively large plastic strains need to occur at the extreme fibres before the full plastic moment can develop. This will tend to limit the maximum rotational capacity of the plastic hinge.

Second, very large member and weld thicknesses result in the development of significant triaxial states of stresses, which have been shown to limit material ductility, even in highly ductile parent material.

9.4   Materials specification

Modern ductile steels produced for non-seismic environments are generally suitable for earthquake resistance, and Eurocode 8 places no additional requirements on basic steel material specification to those of the non-seismic steel Eurocode 3. In US practice, however, certain steel types are excluded from seismic-resisting structures by the AISC standard, and minimum Charpy notch toughness values are also specified where the steel thickness exceeds 30 mm.

The major difference in specification for seismic applications involves the specification of upper bounds on the yield strength. This is to ensure that, during a severe earthquake, parts of the structure designed to yield (the 'dissipative' members, in Eurocode 8 parlance) do not have strengths much greater than their nominal design strength. If they did, brittle elements with yields closer to nominal values might reach their failure strength before the intended yield mechanism formed, thus greatly reducing the available ductility. Eurocode 8 treats this issue by offering the designer three options.

(1) By requiring an upper bound on the yield strength of the dissipative members to be less than a given factor (recommended as 1.375) times the nominal yield strength.
(2) By using a lower grade of steel in the dissipative members than the brittle, or non-dissipative, ones.
(3) For existing structures, by basing the capacity design checks, that brittle elements do not reach their failure strength, on measurements of the actual yield strength of the dissipative elements, as measure *in situ*.

Since the failures in Northridge and Kobe, there has been greater recognition of the importance of specifying ductile welding materials and procedures. Low hydrogen weld metals with good notch ductility are needed, and it is important to ensure that the welding sticks are kept dry before use. Eurocode 8 has no special requirements for specifying weld material, although AISC specifies a minimum Charpy V-notch toughness.

Further advice on material specification for steel members and weld material is given in FEMA 353 (FEMA 2000a).

9.5 Analysis of steelwork structures

Steelwork structures can be analysed by any of the methods described in Chapter 3. Some particular aspects which apply to steel are discussed in the following subsections.

9.5.1 Ductility reduction factors

In linear elastic methods of analysis, relatively large ductility reduction factors are permitted by codes for the most ductile configurations. In Eurocode 8, a behaviour factor q of typically 6 to 8 may be applied to specially detailed moment-resisting and eccentrically braced frames, and the maximum value of the equivalent R factor in the US International Building Code (IBC: ICC 2003) is 8. Lower values apply to concentrically braced frames, typically $q = 4$ and $R = 6$ for the most ductile arrangements. The recommended minimum factors for steel structures without seismic detailing are $q = 1.5$ in Eurocode 8 (compared to 1.5 for reinforced concrete and masonry) and $R = 3.5$ in IBC (compared to 3 for reinforced concrete and 1.5 for plain masonry), although both codes specify that non-seismically detailed structures should not be used in seismic regions, unless they are seismically isolated.

9.5.2 Rotational demand and capacity of steel flexural hinges

If a non-linear static or dynamic analysis is performed, direct information is obtained on the plastic demands in the yielding regions. In principle, this allows a much more rigorous assessment of the ultimate performance of a structure than by use of linear analysis with crude ductility factors such as q in Eurocode 8 or R in IBC. However, assessing the rotational capacity of steel plastic hinges from first principles is much more complex than is the case for reinforced concrete members, primarily because of the effect of local and global buckling on response. In practice, empirical expressions for plastic hinge rotation are needed; data are given in FEMA 356 (FEMA 2000b) and in Annex B of Eurocode 8 Part 3. For design, both Eurocode 8 and AISC set minimum limits on rotational capacity which must be demonstrated directly by testing, unless standard, pre-qualified details are used, as discussed in subsection 9.10.6.

9.5.3 Allowing for flexibility in unbraced steel frames

Two points are worth noting in this context. First, since moment-resisting steel frames are relatively flexible, P–delta effects (Fig. 3.14) may be significant in tall unbraced structures. Second, the flexibility of the panel zones, where beams and columns intersect, may contribute significantly to the deflection of a

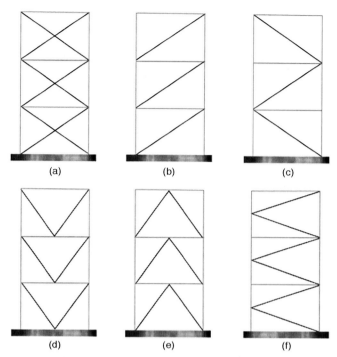

Fig. 9.9 Examples of bracing arrangements for CBFs: (a) X-braced; (b) diagonally braced (example 1); (c) diagonally braced (example 2); (d) V-braced; (e) inverted V-braced; and (f) K-braced. (Note: K-bracing – arrangement (f) – is not permitted for seismic resistance)

moment-resisting steel frame, and should be allowed for. Bruneau *et al.* (1998) provide an extensive discussion of how to allow for this effect in an analysis.

9.5.4 Analysis of concentrically braced frames

When using linear analysis methods, X-braced frames (Fig. 9.9(a)) are generally analysed neglecting the stiffness contribution of the compression members, and the tension members are sized on the same assumption. It is advisable to neglect the contribution of the diagonal braces to resisting gravity loads in any bracing configuration, and also to supporting the horizontal beams in the V-braced configurations of Figs 9.9(d) and 9.9(e).

9.6 Design of steel building structures

The remaining sections of this chapter discuss design considerations for steel structures, with particular reference to the provisions of Eurocode 8 and AISC. For a more detailed discussion of the AISC, refer to Uang *et al.* (2001) and Hamburger and Nazir (2003).

9.7   Design levels of ductility

In both Eurocode 8 and AISC, different levels of ductility are recognised, and here the discussion in section 8.6 for concrete applies equally to steel structures. The ductility classes in Eurocode 8 are high (DCH), medium (DCM) and low (DCL) while the (non-equivalent) ductility levels in AISC are special, intermediate and ordinary.

9.8   Concentrically braced frames (CBFs)

General planning considerations for CBFs were given in subsection 5.4.4 of Chapter 5. Some of the more important design requirements given in Eurocode 8 are as follows. Somewhat different rules apply in AISC, as discussed by Elghazouli (2003).

9.8.1   General

(a) Choice of dissipative elements

Generally speaking, tension diagonals are designed as the 'dissipative' or yielding elements, and other elements, including beams, columns and connections, must be designed on capacity design principles to resist the yield force from these diagonals with sufficient overstrength. In particular, connections must be designed to resist a load equal to 1.375 times the yield strength of the members they connect. This applies to bolted or fillet welded connections, but full-strength butt welds are considered to satisfy the overstrength condition without need for further analysis. Connections may alternatively be designed as dissipative elements, but in this case special design and analysis procedures are required.

(b) Compactness of section

In ductility class high (DCH) structures, bracing members must belong to compactness Class 1 (see subsection 9.3.3); Class 2 members are also permitted in ductility class medium (DCM) structures.

(c) Effect of bracing

Bracing systems which rely on both compressive and tensile braces to resist lateral load (e.g. the V-braced arrangements of Figs 9.9(d) and 9.9(e)) qualify for a much less favourable q factor (2.5 and 2 for DCH and DCM respectively) than X-braced systems (e.g. Fig. 9.9(a)). The latter have $q = 4$ for both DCH and DCM. Note that the K-braced arrangement of Fig. 9.9(f) is not permitted.

(d) Arrangement of tension and compression braces

Within any plane of bracing, the compression diagonal braces should balance the tension diagonal braces at each bracing level, in order to avoid tension braces contributing most to lateral resistance in one direction and compression braces in the other. This is to satisfy the general principle that the diagonal elements of bracings should be placed in such a way that the load–deflection characteristics of the structure are the same for both positive and negative phases of the loading cycle.

(e) Capacity design of columns

In principle, capacity design procedures are required to ensure that the columns can sustain the yield strength of the braces without buckling. In practice, this raises a difficulty, because a strict interpretation would then require the lower columns to be designed for the simultaneous yielding of all bracing higher in the structure, a condition which is unlikely to occur. The implementation in Eurocode 8 (and AISC) therefore requires the columns to be designed for the force obtained from the seismic analysis (i.e. not directly from capacity design principles) but increased by a suitable factor to ensure the columns remain essentially elastic. The factor in Eurocode 8 is around 1.375 times the minimum ratio of actual to required strength in the bracing; the inclusion of the actual to required strength ratio represents a direct application of capacity design principles.

(f) Distribution of ductility demand in braces

It is important to ensure a reasonably uniform distribution of ductility demand in the braces over the height of the structure. If this is not achieved, and the braces at one level yield well before the others, a weak storey might form, concentrating most of the ductility demand at that level. To avoid this, Eurocode 8 places a restriction on the ratio of bracing member strength to strength required from the seismic design. The ratio between maximum and minimum values of this ratio must not exceed 125%. There is no similar requirement in AISC.

9.8.2 X-braced systems

These are generally designed assuming that the compression braces do not contribute stiffness or strength. Eurocode 8 places upper and lower limits on the slenderness of diagonal braces in X-braced systems. The upper limit corresponds to a slenderness l/r_y of around 180 (depending on yield strength), and is designed to prevent the strength and stiffness degradation shown for a slender strut in Fig. 9.4. The lower limit of around 110 is intended to prevent column overloading; columns to which the diagonal braces are connected will be sized to resist the full yield strength of the tension brace assuming no force in the compression brace, but higher axial forces might occur in the columns before very stocky braces have buckled. In AISC, there is a similar limit on upper bound slenderness, but no lower limit.

9.8.3 Diagonal and V-braced systems

These systems rely on both compression and tension braces for stability, and so the stiffness and strength of the compression braces must be explicitly accounted for. The same upper bound limits on slenderness apply, but there is no lower bound limit in Eurocode 8, because the concern about neglecting the compression brace force does not apply.

In V-braced systems, the horizontal brace is subjected to an out-of-balance force when the compression brace begins to buckle, and in Eurocode 8 this must be designed for. Also, the horizontal brace must be designed to carry any gravity loads without support from the diagonal braces, and the AISC rules are similar.

9.8.4   K-braced systems

These are not permitted, because buckling of the compression brace imposes an out-of-balance force not on the horizontal beam (as in the case of V-braced systems) but on the column, and this is clearly unacceptable.

9.9   Eccentrically braced frames (EBFs)

In EBFs, the joint's diagonal bracing members are deliberately separated from those of the vertical and horizontal members to form a link element that can act as a ductile fuse under extreme lateral loads. General planning considerations for EBFs were given in subsection 5.4.5 of Chapter 5. Design rules based on the research effort of the last 30 years (mainly conducted in the USA) are contained in codes such as Eurocode 8 and AISC; they are intended to ensure that the link elements have sufficient ductility to sustain the inelastic cyclic deflections to which they would be subjected in a design earthquake, and to ensure that the surrounding members always remain within the elastic range. The latter objective is met by the use of capacity design procedures – that is, designing the surrounding members for the yielding actions developed in the links. EBFs are assigned ductility factors similar to those of ductile moment frames – that is, a q of up to 8 in Eurocode 8 and an R of 8 in AISC.

Links can be classified as 'short', 'intermediate' or 'long'. Short links yield first in shear along their entire length, while long links yield first in flexure at each end. Intermediate links dissipate energy in both flexure and shear. For a given global ductility demand, short links experience a much greater rotation and hence local ductility demand than long links (Fig. 9.10). However, the plastic strains in short links are spread over the entire length, rather than being concentrated at the ends as they are in long links, so the plastic rotation capacity of short links is around three times greater than that of long links. Moreover, they offer the advantage of providing much greater stiffness before yielding, and hence protection in moderate earthquakes; short links are therefore generally favoured over

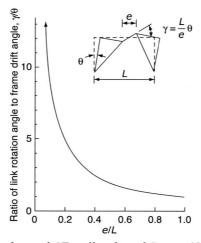

Fig. 9.10   Link rotation demand (Engelhardt and Popov 1989)

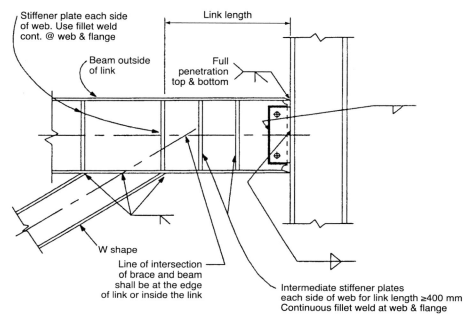

Fig. 9.11 Typical detail of a short link (from AISC 2002)

long ones although all types of links (short, intermediate and long) are permitted in both Eurocode 8 and AISC.

Short links yield in shear and hence are subject to shear buckling of their webs. This needs to be controlled by the provision of vertical web stiffeners welded between the link's flanges, and codes give rules for the required provision of stiffeners. Figure 9.11 shows a typical short link detail from US practice.

Long links are subject to lateral torsional or local flange buckling instability. Lateral restraint is needed to control the former, while a compact section (see subsection 9.3.3) limits the latter. Vertical web stiffeners may also have some limited effect, but are not required by Eurocode 8 or AISC for long links. Full-depth web stiffeners are however required where a diagonal brace meets the link.

The design procedure for EBFs in both Eurocode 8 and AISC is similar. For an elastic analysis, the links are sized from the actions obtained from the seismic analysis, using the specified q or R factor. They can then be classified on the basis of their length e, shear strength V_p and flexural strength M_p as follows.

Short links	$e \leq (1.6M_p)/V_p$
Intermediate links	$(1.6M_p)/V_p < e < (2.5M_p)/V_p$
Long links	$e \geq (2.5M_p)/V_p$

The link classification determines both the rotational capacity of the link and the requirement for web stiffeners. The design rotational capacity (given by Eurocode 8 as 0.08 radians for short links, 0.02 radians for long links, with intermediate links interpolated) must be compared with the rotation demand. In an elastic analysis procedure, the storey drifts (interstorey deflections) are calculated in the normal way, remembering that the deflections from a ductility-modified response spectrum analysis must be factored by the ductility factor (see equation (3.13) in Chapter 3). Thus, in Eurocode 8, the deflections are taken from the design (ductility-modified) spectrum, and factored by q; the procedure in AISC is slightly different in detail though similar in intent. The link rotation is then calculated assuming that all the deformation occurs in the link, the rest of the structure remaining rigid (see for example Fig. 9.10).

Finally, the columns, beams, diagonal braces and connections are sized to resist the actions induced by the yielding link, with a suitable overstrength factor. Details of the procedures are slightly different between AISC and Eurocode 8, but the principles are the same.

9.10   Moment-resisting frames

General characteristics of moment-resisting (unbraced) frames were discussed in subsection 5.4.3 of Chapter 5.

9.10.1   General considerations

The design intent is normally to limit yielding to flexural hinges at the ends of beams, and to ensure that columns remain elastic by the use of capacity design procedures. To this end, Eurocode 8 requires that the flexural strength of columns at a joint exceeds the flexural strength of the beams at the joint by 30% (20% in AISC), except in single-storey frames and in the top storey of multi-storey frames, where the requirement does not apply. However, if the columns are fixed against rotation at their base, plastic hinges must also form there if a sway mechanism is to develop (see Fig. 3.28(a) in Chapter 3) and the columns must be able to sustain the plastic rotations involved. Alternatively, the column bases must be pinned; however, this substantially increases deflections to an extent that may be unacceptable. Given the poor performance of column bases in the Northridge and Kobe earthquakes noted by Smith (1996), attention to the design of column bases is required whatever solution is adopted.

Note that, unlike the case of reinforced concrete frames, a capacity design check for shear strength is not required, because yielding in shear (at any rate for reasonably compact sections without excessively thin webs) is a relatively ductile mechanism. However, column buckling under axial loading is a highly undesirable mechanism, and should be avoided by methods discussed in subsection 9.10.4 below.

9.10.2   Preliminary sizing

A rough preliminary indication of required member sizes may be obtained for a typical building by assuming a total seismic weight of say 10 kN per square metre of floor area (to allow for the structure, permanent finishes and a proportion of the live load). Equivalent lateral forces can then be obtained, using standard code procedures, and the bending forces in the beams estimated by assuming points of contraflexure exist at the midspan of beams and mid-height of columns and that inner columns take twice the shear of external columns. The required beam flexural strengths can then be checked. The column must be sized by capacity design principles; their bending strength should be 30% greater than that of the beams, while simultaneously resisting axial loads due to gravity, plus those induced by flexural yielding of the beams which, for a preliminary design, can conservatively be assumed to take place simultaneously in all beams.

However, deflections rather than strength may well govern. A preliminary estimate of storey drift can be obtained from the equation

$$d = x \left(\frac{k_b + k_c}{k_b k_c} \right) \left(\frac{h^2}{12E} \right) V_c \qquad (9.1)$$

where x is the ductility factor; d is the storey drift (m); $k_b = (I_b/L)$ for a representative beam (m^3); $k_c = (I_c/h)$ for a representative internal column (m^3); I_b, I_c are moments of inertia of beam and column respectively (m^4); L is the centre-to-centre spacing of columns (m); h is the storey height (m); E is Young's modulus of steel (kPa); and V_c is the shear in the representative column (kN).

The ductility factor x is the factor by which the deflections obtained from an elastic analysis must be multiplied to allow for plastic deformations; in Eurocode 8, x is taken as the behaviour factor q, and in IBC it is the factor C_d given in Table 1617.6.2 in the IBC.

The storey drift must then be compared with the maximum permitted in the governing code. In Eurocode 8, this would generally be 1% of the storey height under the ultimate design earthquake, but up to twice this deflection is allowed where the cladding and partitions are not brittle, or are suitably isolated from the frame. IBC generally requires a limit of 1% of the storey height.

A procedure such as this can form the basis for a more rigorous design, perhaps using a computer program, by selecting member sizes that will allow a satisfactory solution to be found without too many trial iterations.

9.10.3 Beams

As described in subsection 9.3.3, the beam's flange and web thickness must be sufficient to limit local flange and web buckling, and Eurocode 8 requires sections that conform to Eurocode 3 Class 1 (or Class 2 for DCM structures) specifications. Lateral torsional buckling must also be controlled by adequate lateral restraint.

Beam flexural strength is assessed on the basis of the seismic analysis, although as noted earlier this may not provide sufficient stiffness to limit deflections to within code limits in tall buildings.

Flexural hinges forming in beams must be capable of sustaining an adequate plastic rotation. A minimum rotation capacity of 0.025 radians for DCM and 0.035 radians for DCH is specified by Eurocode 8, and in addition the loss of stiffness and strength under an unspecified number of cyclic loads should not exceed 20%. In the AISC code, the plastic rotation capacity of special moment frames must be at least 0.030 radians, reducing to 0.020 radians for intermediate ductility frames and 0.010 radians for ordinary frames. In both the AISC and Eurocode, the capacity must be based on testing, and not calculation; either standard pre-qualified designs should be used (see FEMA 350; FEMA 2000c) or special testing of at least two specimens carried out for non-standard designs. Achievement of these rotational capacities is highly dependent on the connection design, as discussed in subsection 9.10.6 below.

9.10.4   Columns

Columns are generally designed to be protected against yielding and therefore do not have to conform to the same compactness requirements referred to above for beams. The exception (noted in subsection 9.10.1 above) is fixed column bases; here, the seismic compactness rules must apply because the columns are 'dissipative'.

Capacity design procedures are used to ensure adequate flexural strength (see subsection 9.10.2). It might be thought they would also be suitable to set design axial loads in columns, to prevent highly undesirable axial buckling failure. In principle, this can be done by adding the axial loads generated by simultaneous yielding of all beams to the gravity loads. However, this simple addition would result in excessive conservatism, in the same way as it would for columns in braced frames (see subsection 9.8.1(e)). This is because it is unlikely that all the beams will yield simultaneously. Generally, codes specify that the columns are designed for the forces derived from the seismic analysis, increased by a simple factor. In Eurocode 8, the factor equals 1.375Ω. Here, Ω is the minimum ratio of resistance moment to design moment at plastic hinge positions in the beams. Since the resistance moment must not be less than the design moment, Ω is always at least 1. Where the beams are all sized considerably in excess of the minimum requirement, the structure will start to yield at a lateral force consider- ably greater than that effectively assumed in the analysis. Therefore, the column forces will also be greater than predicted by the analysis. Factoring the column forces by 1.375Ω therefore allows for this. The factor is set at 1.375Ω rather than Ω to allow for strain hardening in the beam plastic hinges, and to provide some degree of additional reserve of strength.

Column splices need to be designed to transmit safely the design axial force in the column, calculated as above, together with the column bending moment at the splice position. An elastic analysis is a potentially unsafe way of predicting the latter, both because the beams may be stronger than required by the analysis, and so yield at higher loads, and also because the point of contraflexure in the column under inelastic dynamic loading tends to be poorly predicted by an elastic analysis. At the extreme, if the splice were placed at the point of contraflexure (zero moment) predicted by an elastic analysis, and were designed only for axial load, it could fail if the point of contraflexure shifts. Eurocode 8 has no specific requirement to ensure the safe design of column splices, but AISC requires that splices in special moment frames should be designed to develop the expected flexural strength of the smaller of the two columns being connected.

9.10.5   Panel zones

The panel zones of columns are the parts to which the beams connect. They are equivalent to beam–column joints in concrete frames, discussed in subsection 8.7.4 of Chapter 8, and (in moment frames subject to lateral loads) are subject to very high shear forces, as shown in Fig. 8.15. Although yielding of the panel zone in shear is a ductile failure mode, it significantly reduces the stiffness of the frame and is generally discouraged. Panel zones therefore need to be designed to withstand the shear forces induced in them when the beams connected to them yield, by capacity design principles; both plastic yielding and buckling need to be considered. The panel zone shear may be approximated by the following equation

$$V_{pz} \approx \mathbf{f}_y(t_{f1}b_1 + t_{f2}b_2) - (V_3 + V_4)/2 \qquad (9.2)$$

where $\mathbf{f}_y$ is the yield strength of steel; t_{f1} and b_1 are, respectively, the flange thickness and width of beam 1 on one side of the column; t_{f2} and b_2 are, respectively, the flange thickness and width of beam 2 on the other side of the column; V_3 is the shear force in column above the joint; and V_4 is the shear force in column below the joint.

The basis of this equation is that $(\mathbf{f}_y t_f b)$ represents the yield force in the beam flanges, and the equation follows directly from equilibrium considerations.

Panel zones also require detailing. Horizontal stiffeners are needed across the top and bottom of the zone to transmit the beam flange forces into the zone, since these are what give rise to the high shears in the zone. A minimum plate thickness of the panel zone is also advisable to control inelastic web buckling, even if the zone has been designed not to yield on capacity design principles. Bruneau *et al.* (1998) advise a minimum panel zone thickness t_z given by

$$t_z = (d_z + w_z)/90 \qquad (9.3)$$

where d_z is the panel zone depth and w_z is the width between column flanges.

9.10.6 Connections

(a) General
Beam-to-column connections are generally designed on capacity design principles to withstand the yielding forces in the beams that they connect; Eurocode 8 requires design for moments and shears generated by 1.375 times the yield moments in the beams. However, Eurocode 8 does admit the possibility of 'dissipative' connections which are designed to yield and dissipate energy. If such dissipative connections are used, special design procedures are required.

(b) Welding
Since the connection failures in the Northridge earthquake, US codes have not provided design rules for welded beam-to-column connections in moment frames. Instead, connections must be either justified directly by testing, or be based on a pre-qualified standard design, of which examples are given in FEMA 350 (FEMA 2000c). Eurocode 8 also requires experimental justification of welded joints.

Figures 9.12 and 9.13 show two pre-qualified joints from FEMA 350, illustrating two common strategies for achieving satisfactory joints. In the first arrangement, the beam is provided with flange cover plates, forcing the plastic hinge to occur away from the connection.

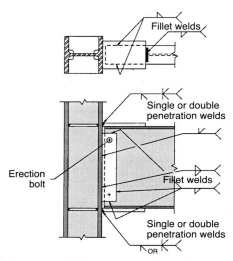

Fig. 9.12 Pre-qualified joint with flange cover plate (from FEMA 350)

In the second arrangement, the beam has been deliberately weakened at a point near the connection, which has the effect of reducing the moments to which the connection is subjected when the beam yields. This arrangement has been used to improve the ductility of existing structures with inadequate, pre-Northridge connections.

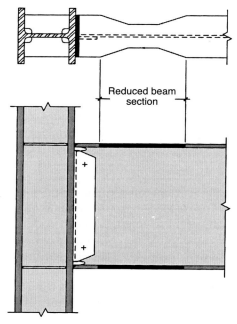

Fig. 9.13 Pre-qualified joint with 'dogbone' (from FEMA 350)

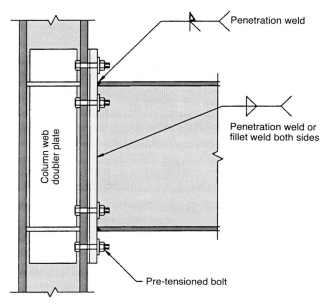

Fig. 9.14   Pre-qualified bolted joint (from FEMA 350)

(c) Bolting

Bolted beam column joints are subject to similar requirements as welded ones. Figure 9.14 shows a pre-qualified bolted joint from FEMA 350; note that the bolts in this case will be subject to significant prying forces which must be accounted for.

Bolted joints may be more suitable for forming partial-strength connections that are designed to dissipate plastic energy. Elghazouli (1998) describes tests carried out on frames with partial-strength joints, and good results are claimed.

9.10.7   Frames not proportioned to resist lateral loads

A common framing plan involves a moment-resisting perimeter frame designed to resist all seismic loads, and internal frames designed only to resist gravity loads. It is of course essential that the latter do not collapse under the deflections to which they are subjected during the design earthquake, and this condition needs to be checked. If the forces in gravity-only frames under the combined action of gravity loads and seismic deflections do not cause their ultimate strength to be exceeded, the condition is satisfied. In fact, gravity-only frames will have some inelastic deformation capacity, so this check is conservative.

9.10.8   Moment-resisting frames with masonry infill panels

Eurocode 8 allows masonry infill panels either to be designed as structurally separated from the steel frame, or to interact with it, in which case the effects of interaction must be considered. For further discussion, see subsection 8.7.8 in Chapter 8 (reinforced concrete design), since the considerations for steel frames are similar to those for concrete frames.

9.11   Steel–concrete composite structures

Steel sections acting compositely with reinforced concrete have certain advantages. The compressive strength added by the concrete increases member resistance to overall buckling, with the possibility of improving cyclic performance. The concrete, if properly detailed, can also control the onset of local flange or web buckling. Eurocode 8 devotes a chapter to steel–concrete composite structures, and they are also covered in the AISC and Japanese seismic design rules. Further information is given by Elnashai *et al.* (1995b).

9.12   References

AISC (2002). *Seismic provisions for structural steel buildings.* American Institute of Steel Construction, Chicago. Free download from www.aisc.org.

Bruneau M., Uang C. M. & Whittaker A. (1998). *Ductile Design of Steel Structures.* McGraw Hill, New York.

CEN (2004). Eurocode 8: EN 1998-1. *Design of structures for earthquake resistance.* Part 1. General rules, seismic actions and rules for buildings. Section 6: Special rules for steel buildings. Section 7: Special rules for composite steel-concrete buildings. European Committee for Standardisation, Brussels.

CEN (2005). Eurocode 3: EN 1993-1-1. *Design of steel structures.* Part 1-1: General rules and rules for buildings. European Committee for Standardisation, Brussels.

EEFIT (1986). *The Mexican Earthquake of 19th September 1985.* Society for Earthquake and Civil Engineering Dynamics (SECED), London.

Elghazouli A. (1998). Seismic design of steel frames with partial strength connections. In: Booth E. (ed.) *Seismic Design Practice into the Next Century: Research and Application.* A A Balkema, Rotterdam.

Elghazouli A. (2003). Seismic design procedures for concentrically braced frames. *Structures and Buildings,* Vol. 156, November, pp. 381–394.

Elnashai A. S., Bommer J., Baron C., Lee D. & Salama A. (1995a). *Selected Engineering Seismology and Structural Engineering Studies of the Hyogo-ken Nanbu (Great Hanshin) Earthquake of 17th January 1995.* ESEE research report No. 95-2, Civil Engineering Department, Imperial College, London.

Elnashai A. S., Broderick B. M. & Dowling P. J. (1995b). Earthquake-resistant composite steel/concrete structures. *The Structural Engineer,* Vol. 73, No. 8, pp. 121–132.

Engelhardt M. & Popov E. (1989). *Behaviour of long links in eccentrically braced frames.* Report No. UCB/EERC-89/01. Berkeley Earthquake Engineering Research Center, University of California.

Engelhardt M. & Sabol T. (1996). Lessons learnt on steel moment frame performance from the 1994 Northridge earthquake. In: Burdekin M. (ed.) *Seismic design of steel buildings after Kobe and Northridge.* Institution of Structural Engineers, London.

FEMA (2000a). FEMA 353. *Recommended specifications and quality assurance guidelines for steel moment–frame construction for seismic applications.* Federal Emergency Management Agency, Washington DC.

FEMA (2000b). FEMA 356. *Prestandard and commentary for the seismic rehabilitation of buildings.* Prepared by American Society of Civil Engineers for Federal Emergency Management Agency, Washington DC.

FEMA (2000c). FEMA 350. *Recommended design criteria for new steel moment–frame buildings.* Federal Emergency Management Agency, Washington DC.

Hamburger R. & Nazir N. (2003). Seismic design of steel structures. In: Chen W.-F. & Scawthorn C. (eds) *Earthquake Engineering Handbook.* CRC Press, Boca Raton, FA.

ICC (2003). IBC: 2003. *International Building Code*. International Code Council, Falls Church, VA.

Jain A. K., Goel S. C. & Hanson R. D. (1978). *Hysteresis behaviour of bracing members with different proportions*. Report No. UMEE 78RE July. Ann Abor: Department of Civil Engineering. The University of Michigan.

Smith D. G. E. (1996). An analysis of the performance of steel structures in earthquakes since 1957. In: Burdekin M. (ed.) *Seismic Design of Steel Buildings after Kobe and Northridge*. Institution of Structural Engineers, London.

Uang C.-M., Bruneau M., Whittaker A. & Tsai K.-C. (2001). Seismic design of steel structures. In: Naeim F. (ed.) *The Seismic Design Handbook*. Kluwer Academic Publishers, Boston.

Vann W. P., Thompson L. E., Whalley L. E. & Ozier L. D. (1973). Cyclic behaviour of rolled steel members. *Proceedings of the 5th World Conference on Earthquake Engineering*, Vol. 1, pp. 1187–1193. International Association for Earthquake Engineering, Rome.

Yanev P., Gillengerten J. D. & Hamburger R. O. (1991). *The Performance of Steel Buildings in Past Earthquakes*. American Iron & Steel Institute, Washington DC.

10 Masonry

'Masonry materials – mortar and stones or bricks – are stiff and brittle, with low tensile strength, and are thus intrinsically not resistant to seismic forces. However, the earthquake resistance of masonry as a composite material can vary between good and poor, depending on the materials used ... [and] ... the quality of workmanship.'

Sir Bernard Feilden. In: *Between Two Earthquakes – Cultural Property in Seismic Zones*. ICCROM, Rome/Getty Conservation Institute, Marina del Rey, CA, 1987

This chapter covers the following topics.

- The lessons from earthquake damage
- Characteristics of masonry as a seismic-resisting material
- Material specification
- Special considerations for analysis
- Masonry walls
- Floors and roofs in masonry buildings
- Masonry as non-structural cladding

10.1 Introduction

Brick and stone masonry is a widely available, low-energy material, and the skills are found all over the world to use them for creating highly practical and often beautiful buildings. However, its low tensile strength limits the available ductility and places reliance on its ability to sustain high compressive stresses during an earthquake. If the compressive strength is low (as is the case for example with earth bricks or 'adobe') then the consequences in an earthquake can be disastrous, and often have been (Fig. 1.9). However, well-designed buildings made from good-quality brick or stone can perform well. In US practice, all new masonry buildings in areas of high seismicity have to be reinforced with steel. By contrast, Eurocode 8 permits the use of unreinforced masonry to withstand strong earthquakes, although it is unlikely that a building taller than one or two storeys could be made to comply with the code if the seismicity is high.

10.2 Forms of masonry construction and their performance in earthquakes

Masonry consists of blocks or bricks, usually bonded with mortar. A wide variety of forms exist. The weakest is where cohesive soil is placed in a mould and sun-dried to form a building block. This type of construction (called adobe in Latin America and elsewhere) is cheap, widely available and requires only basic skills to form, but cannot be relied on to resist strong ground motion. Stabilising the soil with lime or other cementitious material improves matters.

Random rubble masonry consists of rough cut or natural stones held in a matrix of soil or mortar. It may form the core of a wall with a cladding of dressed (i.e. cut) stone, called ashlar. The seismic resistance depends on the matrix holding the stones together; if this is weak, the seismic performance will be poor or very poor.

Carefully cut rectangular blocks of stone (dressed stone) of good quality arranged to resist lateral resistance without developing tensile stresses can possess surprisingly good earthquake resistance. Here, the presence of vertical prestress, usually coming from the weight of masonry above, is important for two reasons. First, seismically induced tensile stresses may not develop if the prestress is great enough. Second, the shear strength of dressed stone relies primarily on friction; the higher the contact forces between stones, the higher the shear strength. Since compressive gravity loads are higher at the base of a building, often the seismic resistance is also greater, and so often the damage observed in dressed stone masonry is less at the bottom of a building than at the top (Fig. 10.1). By contrast,

Fig. 10.1 Increase in seismic damage with height in a stone masonry building, Gujarat, India, 2001

Fig. 10.2   Poor performance of hollow clay tile masonry in Erzincan, Turkey, 1992

the opposite is usually the case for structures in steel and concrete because the highest seismic forces occur at the bottom of the building (as they do in masonry buildings) but the gravity preload is likely to weaken steel and concrete structures, rather than strengthening them as it can do in stone masonry. Inducing compressive stresses by introducing vertical or inclined steel prestressing cables is thus a powerful way to improve the seismic resistance of good-quality stone masonry buildings (see Beckmann and Bowles 2004, section 4.5.10)

Manufactured bricks or blocks can approach the compressive strength of natural stone without requiring the special skills and equipment needed to dress natural stone. They may be reinforced with steel laid in some of the horizontal mortar bed joints (e.g. every third joint) and with vertical reinforced concrete elements, particularly at corners and around openings; this can form a satisfactory seismic resisting system. Hollow clay bricks are lighter but much weaker and have not performed well seismically unless reinforced or confined within a beam–column frame (Fig. 10.2). Concrete hollow blocks, often made with lightweight aggregates, are cast with central voids, which can be reinforced and concreted to form a strong, monolithic system (Fig. 10.3). Proprietary brick systems have

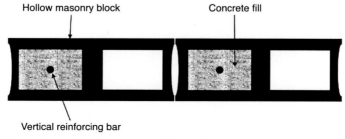

Fig. 10.3   Typical reinforced concrete hollow blocks

also been developed which provide a mechanical interlock between bricks, which improves the shear strength of the completed wall.

10.3   Designing masonry for seismic resistance

10.3.1   Classification of masonry walls

Eurocode 8 classifies masonry walls into one of three types

(1) unreinforced masonry
(2) confined masonry
(3) reinforced masonry.

Unreinforced masonry walls rely solely on the strength of the masonry to resist seismic effects. However, Eurocode 8 specifies that horizontal concrete beams or steel ties should be placed around the building perimeter at every floor level with a minimum steel area of $200\,\text{mm}^2$.

In confined masonry, the walls are surrounded by vertical and horizontal reinforced concrete elements. These must be cast into the walls after their construction, ensuring a good bond between the concrete confining elements and the masonry. This type of wall is distinct from masonry infill panels built into a concrete frame after its construction; the latter is treated as a concrete frame structure, rather than a masonry shear wall structure and was discussed in subsection 8.7.8. Vertical elements are needed at minimum at the corners of the building, at the free ends of walls and around openings exceeding $1.5\,\text{m}^2$ in area; in no case should the spacing between vertical elements exceed 5 m. Horizontal elements are required at each floor level (or at 4 m centres, if less) and around openings. Rules are given for the minimum longitudinal and transverse steel required in the confining elements.

In Eurocode 8 (CEN 2004), reinforced masonry is required to have a minimum percentage of 0.05% steel horizontally and 0.08% vertically. As previously mentioned, new masonry construction in areas of high seismicity is required to be reinforced in US practice, a notable difference from the Eurocode. Table 10.1 summarises some requirements for steel in special reinforced masonry buildings in seismic regions given by the US code ACI 530 (ACI 2002).

Table 10.1 Minimum steel requirements in ACI 530 for special reinforced masonry buildings

Maximum horizontal and vertical spacing of reinforcing steel	(i) $\dfrac{\text{Length or height of building}}{3}$
	(ii) 1219 mm if less
Minimum area of vertical steel	One third minimum required area of horizontal shear steel
Horizontal shear reinforcement must be anchored around vertical reinforcement with a standard hook	

10.3.2 Minimum material strength

Eurocode 8 recommends a minimum compressive strength of block or brick of $5\,\text{N/mm}^2$ normal to the bed face and $5\,\text{N/mm}^2$ parallel to the bed face in the plane of the wall. A minimum mortar strength of $5\,\text{N/mm}^2$ is recommended for unreinforced and confined masonry, and $10\,\text{N/mm}^2$ for reinforced masonry.

These limits are appropriate for new construction. Existing buildings with lime mortars are unlikely to comply, however, although it may still be possible to demonstrate satisfactory seismic performance. Generally, it is desirable for shear failure to occur in the mortar before failure in the masonry unit, because a more ductile response is likely, mobilising frictional resistance.

10.3.3 In-plane shear strength of masonry walls

In simple masonry wall buildings, practically all the lateral resistance is provided by the in-plane stiffness of the walls. The designer's task is then to ensure that all the seismic forces can be safely transmitted back to the walls by the floors and roof, and to ensure that the walls have sufficient in-plane shear strength to resist them. In most cases, the in-plane shear strength will be governed by the mortar, but with weak stones or bricks and strong mortar, the masonry blocks may fail first (a less ductile mode, as noted previously). Where the mortar shear strength governs, in-plane shear strength is determined by adding the shear resistance of the masonry under zero compression to the shear resistance provided by friction between the masonry blocks. Eurocode 8 specifies that the design seismic in-plane shear strength should be determined from Eurocode 6 (CEN 2005), which provides the following equations. Equation (10.1) applies to masonry where the vertical (header) joints are completely filled with mortar. The limit in equation (10.2) represents failure of the masonry units before slip develops in the mortar.

$$v_\text{d} = \qquad (v \qquad + \qquad 0.4\sigma_\text{v})/\gamma_\text{m} \qquad\qquad (10.1)$$

$$\quad\;\;\text{Intrinsic shear strength}\quad\text{Frictional component}$$

or

$$0.065f_\text{b}/\gamma_\text{m} \quad \text{if less} \qquad\qquad\qquad\qquad (10.2)$$

where v_d is the design in-plane shear strength, v is the masonry shear strength under zero compressive load, σ_v is the vertical stress due to permanent loads, and f_b is the compressive strength.

Note that v_d is related to the average shear stress over the wall, and allows for the fact that the shear stresses in the centre of the wall are greater than those at its ends. Typical values of v for unreinforced masonry are given in Table 10.2. The material factor γ_m is at minimum 1.5. In existing buildings, v may be determined from in-situ tests, such as UBC 21-6 (ICBO 1997). In this test, both header joints of a selected brick are cleared and a flat jack introduced into one of them to stress the brick horizontally in shear. The shear force and deflection at first slip and at failure are noted. Ideally, the force–deflection characteristics should be measured at various levels of vertical load by introducing a second jack to vary the vertical stress in the brick being tested. This enables the frictional component of resistance and coefficient of friction to be estimated.

Table 10.2 Indicative values for intrinsic shear strength of unreinforced masonry

Masonry quality	v: MPa
Poor	0.15
Average	0.30
Good	0.45

In the Eurocode 6 formula quoted in equation (10.1), the frictional component assumes a coefficient of friction of 0.4; in US practice, the coefficient of friction is taken as 1 (FEMA 356; FEMA 2000).

Reinforced masonry can achieve much higher shear strength. Not only does the reinforcement increase the shear strength, but also provides a measure of ductility, as reflected in the q factors shown in Table 10.4.

Door and window openings will of course reduce in-plane shear strength. The shear strength of a wall should be based on its net area, after allowing for openings, and timber or reinforced concrete lintels placed over the openings. Preferably the vertical sides and bottom should be similarly reinforced, and this is a requirement in Eurocode 8 for openings greater than $1.5\,\text{m}^2$ in confined masonry.

Unreinforced masonry walls are likely to suffer a considerable loss of in-plane shear strength and stiffness once their ultimate strength is reached. If there are adjacent, less highly stressed walls, these may then be able to relieve some of the loads, provided the floor or roof diaphragms are strong and stiff enough for the redistribution of forces involved. Eurocode 8 allows for up to 25% of seismic loads to be redistributed from more highly to less highly stressed walls, provided the diaphragms can make the necessary transfers. However, walls that are heavily damaged by in-plane shear forces are likely to have their ability to carry gravity loads compromised as well; hence the likelihood of collapse is increased. For this reason, Eurocode 8 recommends that unreinforced masonry walls should be designed for a q factor of 1.5, reflecting an essentially elastic response (Table 10.4). The improved ability of confined and reinforced masonry to maintain vertical resistance after sustaining significant in-plane shears is reflected in the higher q factors shown in Table 10.4.

Slender masonry walls with a low ratio of base length l to height h may start to uplift at one edge before their in-plane shear strength is reached. This constitutes rocking; although a wall will maintain some shear resistance after rocking has started, the shear stiffness will drop considerably, and it is likely that most of its shear load will shed to other walls. Eurocode 8 specifies that slender walls not satisfying the minimum (l/h) ratios shown in Table 10.3 should be taken as secondary elements which are not counted as contributing to lateral resistance.

10.3.4 Out-of-plane strength of masonry walls

The main task for masonry walls is to resist the overall seismic shear forces developed in the direction in which they run. However, the walls will also be subjected to seismic accelerations perpendicular to their plane, and these will

Table 10.3 Geometric limits on masonry walls from Eurocode 8

Masonry type	Min. thickness, $t_{ef,min}$: mm	Max. out-of-plane slenderness, $(h_{ef}/t_{ef})_{max}$	Min, stockiness, $(l/h)_{min}$
Unreinforced, with natural stone units	350	9	0.5
Unreinforced, with any other type of units	240	12	0.4
Unreinforced, with any other type of units, in cases of low seismicity	170	15	0.35
Confined masonry	240	15	0.3
Reinforced masonry	240	15	No restriction

t_{ef} = thickness of the wall (see EN 1996-1-1: 2005; CEN (2005));
h_{ef} = effective height of the wall (see EN 1996-1-1: 2005; CEN (2005));
h = greater clear height of the openings adjacent to the wall;
l = length of the wall.

give rise to out-of-plane shears. These arise only from the self-mass of the walls and any finishes applied to them, but the shears must still be transferred back to the points of lateral restraint to the wall, and this may involve the development of significant out-of-plane bending moments. Its tensile strength means that this is unlikely to be a problem for reinforced masonry. However, in unreinforced (and to a lesser extent confined) masonry, the very low tensile strength implies that out-of-plane bending strength relies on the compressive prestress due to the gravity loads the walls support. Clearly, out-of-plane bending is likely to be most significant if the wall thickness is small in relation to the distance to lateral restraint. The problem is complex, but several factors make it less severe than it first appears. Tests at Bristol University (Zarnic *et al.* 1998) showed that compressive membrane action can considerably improve the out-of-plane resistance of wall panels. Moreover, once the wall has cracked in out-of-plane bending, it loses stiffness and this is likely to decouple it from the input motions. Unlike in-plane effects, which involve tributary inertia forces from the entire building, out-of-plane response is driven only by self-mass, and hence is likely to be displacement limited. However, excessively small ratios of thickness to storey height or wall length should be avoided; if the out-of-plane deflection of a load-bearing wall takes the line of action of the gravity loads much beyond the central third of the wall thickness, its collapse becomes much more likely. Suitable geometric limits are given in Table 10.3.

Where the lateral restraint from cross-walls and floors is ineffective, out-of-plane failure may also occur by an overall overturning of the wall (Figs 10.4 and 10.5); D'Ayala and Speranza (2003) provide a detailed discussion. This type of failure mechanism, commonly found in historic masonry buildings, points to the importance of ensuring that adequate lateral restraint is provided, for example by providing horizontal tying bands around the building and improving connections between floor and walls.

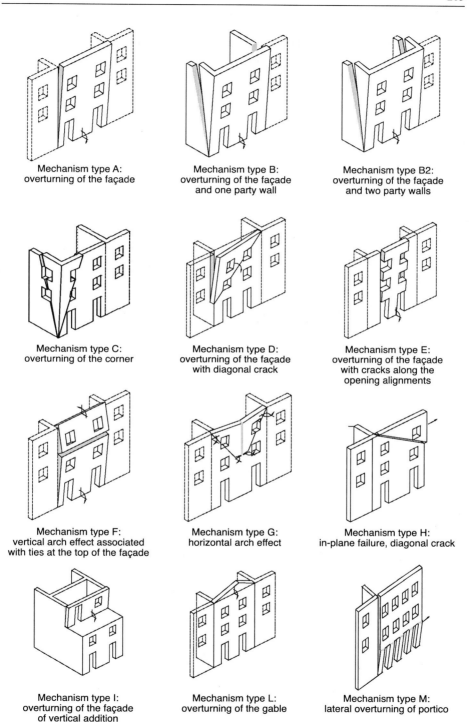

Mechanism type A:
overturning of the façade

Mechanism type B:
overturning of the façade
and one party wall

Mechanism type B2:
overturning of the façade
and two party walls

Mechanism type C:
overturning of the corner

Mechanism type D:
overturning of the façade
with diagonal crack

Mechanism type E:
overturning of the façade
with cracks along the
opening alignments

Mechanism type F:
vertical arch effect associated
with ties at the top of the façade

Mechanism type G:
horizontal arch effect

Mechanism type H:
in-plane failure, diagonal crack

Mechanism type I:
overturning of the façade
of vertical addition

Mechanism type L:
overturning of the gable

Mechanism type M:
lateral overturning of portico

Fig. 10.4 Out-of-plane failure mechanisms in walls (D'Ayala and Speranza 2003)

Fig. 10.5 Overturning failure of a façade in a historic masonry building

10.3.5 Other structural systems in masonry

The discussion so far has applied to conventional buildings with vertical masonry walls, and horizontal floors or roof diaphragms in timber or concrete. However, surprisingly good seismic resistance can be obtained by exploiting the high compressive strength of good-quality stone masonry and its high shear strength when the joints are under compression. Moreover, such structures can be very tolerant of large displacements. Some examples of masonry structures that have survived severe earthquakes are shown in Figs 10.6 and 10.7. Codes of practice are not intended to deal with such structures, and help must be sought elsewhere. Feilden (1982, 1987) and Beckmann and Bowles (2004) provide useful advice. Specialist discrete element software (Brookes and Mehrkar-Asl 1998) may provide some analytical insights into such systems, but their response is highly complex and judgement is likely to prove important in interpreting results.

Fig. 10.6   Response of a masonry arch to strong shaking, Gujarat, India earthquake, 2001. Although clearly in an unsafe condition, this arched masonry structure has just managed to remain stable, despite very large deformations

Fig. 10.7   Response of masonry columns supporting a heavy masonry roof slab, Gujarat, India earthquake, 2001. The columns of this ancient structure, which were unreinforced, showed some cracking but survived an earthquake which toppled many modern reinforced buildings in the Indian city of Ahmadabad

10.4   Analysis of masonry structures

Although the response of the walls is likely to become highly non-linear once they have cracked, a linear static or response spectrum analysis using force reduction factors is likely to give sufficient information on these points in well-designed buildings of regular form. Non-linear static or dynamic analysis may be employed to investigate redistribution of forces between walls in existing buildings, where one or more walls reach their ultimate strength for a seismic intensity below that considered for design.

It is commonly recommended that the stiffness of the walls should be taken as half the initial elastic stiffness, to allow for the effect of cracking. Concrete floors can be taken as effectively rigid in-plane, although their attachment to the walls needs to be checked as having adequate shear strength. The flexibility of timber floors may affect the distribution of shear between walls significantly, and so should generally be included.

Table 10.4 shows the force reduction factors specified in Eurocode 8. In IBC (ICC 2003), special reinforced masonry bearing walls qualify for an R factor of 5 and (in zones of high seismicity) are limited in height to 30 m (100 ft).

Further advice on the analysis of masonry buildings is given in Chapter 7 of Paulay and Priestley (1992).

Table 10.4 Seismic reduction factors for masonry walls in Eurocode 8 (CEN 2004)

Type of construction	Behaviour factor q in EC8
Unreinforced masonry in accordance with EN 1996 (CEN 2005) alone (recommended only for low seismicity cases)	1.5
Unreinforced masonry in accordance with EN 1998-1 (CEN 2004)	1.5
Confined masonry	2.0
Reinforced masonry	2.5

10.5   Simple rules for masonry buildings

In conventional masonry wall buildings, the key features to consider are limiting in-plane shear stresses, ensuring a reasonable distribution of walls in both horizontal directions and an absence of torsional eccentricity, providing efficient floor and roof diaphragms to tie the building together and distribute seismic loads back to the walls, and ensuring that door and window openings do not introduce local points of weakness. Rules of thumb for providing sufficient shear resistance may often be just as valuable when justifying a masonry building (if not more so) as explicit analysis. Eurocode 8 provides some rules for 'simple' low-rise masonry buildings which can be used without further quantitative analysis. The main rules concern the minimum area of wall that should be provided in each direction as a percentage of the total floor plan area; the relevant table is reproduced in Table 10.5.

Table 10.5 Rules for minimum area of shear walls for 'simple' masonry buildings, from Eurocode 8

Type of construction	Number of storeys, n (note 2)	Acceleration at site, a_gS			
		$\leq 0.07kg$	$\leq 0.10kg$	$\leq 0.15kg$	$\leq 0.20kg$
		Minimum sum of cross-sectional areas of horizontal shear walls in each direction, as percentage of the total floor area per storey, $p_{A,min}$			
Unreinforced masonry	1	2.0%	2.0%	3.5%	n/a
	2	2.0%	2.5%	5.0%	n/a
	3	3.0%	5.0%	n/a	n/a
	4	5.0%	n/a (note 4)	n/a	n/a
Confined masonry	2	2.0%	2.5%	3.0%	3.5%
	3	2.0%	3.0%	4.0%	n/a
	4	4.0%	5.0%	n/a	n/a
	5	6.0%	n/a	n/a	n/a
Reinforced masonry	2	2.0%	2.0%	2.0%	3.5%
	3	2.0%	2.0%	3.0%	5.0%
	4	3.0%	4.0%	5.0%	n/a
	5	4.0%	5.0%	n/a	n/a

Notes:
1. The table is based on a minimum compressive strength of $12\,N/mm^2$ for unreinforced masonry and $5\,N/mm^2$ for confined and reinforced masonry.
2. Roof space above full storeys is not included in the number of storeys.
3. For buildings where at least 70% of the shear walls under consideration are longer than 2 m, the factor k is given by $k = 1 + (l_{av} - 2)/4 \leq 2$ where l_{av} is the average length, expressed in metres, of the shear walls considered. For other cases $k = 1$.
4. n/a means 'not acceptable'.

There are number of other conditions that have to be met for masonry buildings to qualify as 'simple'.

(*a*) The plan shape must be approximately rectangular, with a recommended minimum ratio of shortest to longest side of 0.25, and with projections or recesses from the rectangular plan area not exceeding 15%.

(*b*) The building should be stiffened by shear walls, arranged almost symmetrically in plan in two orthogonal directions.

(*c*) A minimum of two parallel walls should be placed in two orthogonal directions, the length of each wall being greater than 30% of the length of the building in the direction of the wall under consideration.

(*d*) At least for the walls in one direction, the distance between these walls should be greater than 75% of the length of the building in the other direction.

(*e*) At least 75% of the vertical loads should be supported by the shear walls.

(*f*) Shear walls should be continuous from the top to the bottom of the building.

(*g*) Differences in mass and shear wall area between any two adjacent storeys should not exceed 20%.

(*h*) For unreinforced masonry buildings, walls in one direction should be connected with walls in the orthogonal direction at a maximum spacing of 7 m.

Further simple rules for traditional earth, brick and stone buildings are provided by Coburn *et al.* (1995) and by Patel *et al.* (2001).

10.6   References

ACI (2002). ACI 530. *Building code requirements for masonry structures.* American Concrete Institute, Farmington Hills, MA.

Beckmann P. & Bowles R. (2004). *Structural Aspects of Building Conservation.* Elsevier, Amsterdam.

Brookes C. & Mehrkar-Asl S. (1998). Numerical modelling of masonry using discrete elements. In: Booth E. (ed.) *Seismic Design Practice into the Next Century. Sixth SECED Conference.* Balkema, Rotterdam.

CEN (2004). EN 1998-1: 2004. *Design of structures for earthquake resistance.* Part 1: General rules, seismic actions and rules for buildings. European Committee for Standardisation, Brussels.

CEN (2005). Eurocode 6. EN 1996-1-1. *Design of masonry structures. Common rules for reinforced and unreinforced masonry structures.* European Committee for Standardisation, Brussels.

Coburn A., Hughes R., Pomonis A. & Spence R. (1995). *Technical Principles of Building for Safety.* Intermediate Technology Publications Ltd, London.

D'Ayala D. & Speranza E. (2003). Definitions of collapse mechanisms and seismic vulnerability of historic masonry buildings. *Earthquake Spectra*, Vol. 19, No. 3, pp. 479–510.

Feilden B. (1982). *Conservation of Historic Buildings.* Architectural Press, Oxford.

Feilden B. (1987). *Between Two Earthquakes – Cultural Property in Seismic Zones.* ICCROM, Rome/Getty Conservation Institute, Marina del Rey, CA.

FEMA (2000). FEMA 356. *Prestandard and commentary for the seismic rehabilitation of buildings.* Federal Emergency Management Agency, Washington DC.

ICBO (1997). UBC 21-6. *In place masonry shear tests.* International Conference of Building Officials, Whittier, CA.

ICC (2003). IBC: 2003. *International Building Code.* International Code Council, Falls Church, VA.

Patel D. B., Patel D. K. & Pindoria K. (2001). *Repair and strengthening guide for earthquake damaged low rise domestic buildings in Gujarat, India.* www.arup.com/geotechnics/project.cfm?pageid = 701

Paulay T. & Priestley M. J. N. (1992). *Seismic Design of Reinforced Concrete and Masonry Buildings.* Wiley, New York.

Zarnic R., Gostic S., Severn R. T. & Taylor C. A. (1998). Testing of masonry infilled frame reduced scale building models. In: Booth E. (ed.) *Seismic Design Practice into the Next Century.* Balkema, Rotterdam.

11 Timber

'Wood construction is light, and while there have been some horrible failures, there have been very few casualties.'

Henry Degenkolb. *Connections*. Earthquake Engineering Research Institute, Oakland, CA, 1994

This chapter covers the following topics.

- The lessons from earthquake damage
- Characteristics of timber as a seismic-resisting material
- Material specification
- Codes, standards and design recommendations

11.1 Introduction

Timber is perhaps the least researched and written about of seismic-resisting materials, although the situation has been redressed to some extent after wooden buildings performed worse than expected in the 1994 Northridge earthquake. Timber is however widely used to provide earthquake resistance in highly seismic areas, both in the developed world, particularly California and New Zealand, and also the developing world, particularly Central and South America where timber and bamboo is widely used for low-rise construction. In the ancient world, wooden pagodas and temples have successfully survived earthquakes in Japan and China and the timber buildings of Anatolia, Turkey also have a good record. Timber also has a long and successful history of use as a horizontal and vertical tie within masonry buildings to improve earthquake resistance. As a renewable resource not requiring highly industrialised methods, it has other advantages, particularly if it is supplied from sustainable sources.

11.2 Characteristics of timber as a seismic-resisting building material

Timber has a high ratio of tensile and compression strength to weight, which is a favourable seismic-resisting feature. Timber joints can also dissipate significant amounts of energy when they are stressed in an earthquake; where yielding of steel elements (nails, screws, bolts) within the joint is involved, damping ratios of as much as 45% can occur, and in most timber structures the damping will be at least 15%, which is two or three times the typical level in concrete or steel structures (Dolan 2003). The failure of the parent timber, however, tends to be rather brittle in most failure modes, and so overstrength members and understrength connections are usually indicated. However, compression failure perpendicular to the grain is ductile, involving collapse of the wood's cellular structure, and some of the ductility in nailed and bolted joints arises from this mechanism. Glued joints are not able to dissipate much energy, and nor are joints made with large steel bolts where failure occurs by crushing, shearing or splitting of the timber. 'Carpenter joints' (such as a tenon or halving joint, where the forces are transferred directly through the wood without mechanical fasteners) may be dissipative, provided shear failure or tension failure perpendicular to the grain does not occur.

Three other favourable seismic features of timber housing should be mentioned. First, it is easy to achieve a good tensile strength in the connections between timber elements, and so a timber-frame building with timber floors is well tied together in a way much harder to achieve in unreinforced masonry. This greatly improves its earthquake resistance. Second, timber frames tend to be quite highly redundant, which also improve resistance. Third, it is straightforward to nail plywood wall panels to a timber frame, which provides lateral strength and stiffness together with excellent energy dissipation through the nailed joints. This form of construction has an excellent performance record in Californian earthquakes.

Two unfavourable features should also be noted. First, the strength of timber reduces when its moisture content increases. It is also susceptible to insect and fungal attack, and this can effectively destroy its resistance. Timber treatment, combined with suitable detailing to dissipate moisture and deter insects, is needed to counter this. Second, devastating fires have broken out after earthquakes in dense areas of wooden buildings causing extensive damage and loss of life, notably in Tokyo in 1923, but also more recently for example in the Marina district of San Francisco in 1989 and Kobe Japan in 1995.

11.3   The lessons from earthquake damage

In recent Californian earthquakes, low-rise timber houses have suffered where they have been inadequately anchored to their foundations, and have shifted. Unbraced 'cripple walls' (short walls lifting the lowest floor off ground level to allow the passage of services) have also commonly failed. A number of buildings with garages at ground level failed due to 'soft storey' formation, particularly 'tuck under buildings' with the lowest level open on three sides and closed on the fourth. Generally, however, the seismic performance of low-rise timber-frame buildings in California is assessed as far superior to that of their unreinforced masonry equivalents, and comparable to that of low-rise buildings with reinforced concrete or reinforced masonry shear walls (Anagnos et al. 1995).

Traditional single-family Japanese houses are one- and two-storey wood post and lintel, with bamboo reinforced mud infill and heavy fired clay tiling. They perform very poorly in earthquakes, being prone to pancake collapse, as seen in the 1995 Kobe earthquake. More recent construction has lighter, shingle roofs, with a more substantial timber frame often on a reinforced concrete base, and this has performed much better (Scawthorn et al. 2005).

In developing countries, single or two-storey housing using bamboo or more conventional timber has generally fared better than masonry construction. Often, however, the housing is occupied by the poorest families forced onto marginal and unsuitable land, and the failures that occur are of unstable slopes on which the houses are built, or are due to weakening of the frames by insect or fungal attack.

A detailed survey of the performance of timber buildings in earthquakes is given by Karacabeyli and Popovski (2003).

11.4   Design of timber structures

11.4.1   Provisions of Eurocode 8

The EC8 chapter on the seismic design of timber structures, which is contained in section 8 of Eurocode 8 Part 1 (CEN 2004a), is relatively short, running to six pages compared to the 58 for concrete and 23 for steel. It is mainly concerned with setting out broad principles for successful design, rather than giving detailed design rules. The main points are summarised below.

EC8 recognises three classes of timber structure, depending on the ability to dissipate energy (Table 11.1). Low ductility (DCL) structures must be designed as elastically responding, with a q factor of 1.5, and the partial factors for materials γ_M given in Eurocode 5 (CEN 2004b) for fundamental load combinations apply. Individual countries (in their National Annex) may prohibit the use of DCL structures in areas of high seismicity, although the main part of Eurocode 8 gives no advice on this. Medium and high ductility (DCM and DCH) structures can be designed for q factors as high as 5, and the more favourable γ_M factors for accidental load combinations apply. For all ductility classes, load combinations

Table 11.1 q factors for timber buildings from Eurocode 8

Design concept and ductility class	q	Examples of structures
Low capacity to dissipate energy (DCL)	1.5	Cantilevers; beams; arches with two or three pinned joints; trusses joined with connectors
Medium capacity to dissipate energy (DCM)	2	Glued wall panels with glued diaphragms, connected with nails and bolts; trusses with dowelled and bolted joints; mixed structures consisting of timber framing (resisting the horizontal forces) and non-load-bearing infill
	2.5	Hyperstatic portal frames with dowelled and bolted joints
High capacity to dissipate energy (DCH)	3	Nailed wall panels with glued diaphragms, connected with nails and bolts; trusses with nailed joints
	4	Hyperstatic portal frames with dowelled and bolted joints
	5	Nailed wall panels with nailed diaphragms, connected with nails and bolts

including earthquake is regarded as instantaneous loading, for which the appropriate strength increase may be applied.

At the most fundamental level, the distinction between 'medium' and 'high' ductility may be established by test; thus, DCM structures must survive three fully reversed cycles to a displacement ductility of 4 with a loss of not more than 20% in resistance. The same test applies for DCH, except that the displacement ductility demand rises to 6. In practice, some rules are given to allow design to proceed for straightforward cases without testing. Some basic rules are as follows.

(a) Glued connections must be regarded as non-dissipative although, as can be seen from Table 11.1, structures with glued diaphragms combined with more dissipative connections can be regarded as having medium or high ductility.

(b) 'Carpenter joints' (defined above) can be regarded as dissipative provided they have sufficient overstrength in shear (a factor of 1.3 on required resistance compared with demand is recommended) and do not fail in tension perpendicular to the grain.

(c) Sheathing for shear walls or floor diaphragms consists of particle board with a minimum density of $650\,kg/m^3$ and a minimum thickness of 13 mm, or of plywood at least 9 mm thick.

(d) Blocking (backing timbers) are required in sheathing for shear walls or floor diaphragms at free edges and over supporting walls.

Specific rules for connections are as follows.

(a) Where smooth nails, dowels or staples are used, there must be additional provision (e.g. retaining straps) to prevent their withdrawal. As a matter of general principle, screws are always preferred to nails or staples, although this is not explicitly stated in the code.

(b) In nailed, dowelled or bolted timber-to-timber or timber-to-steel connections, the minimum thickness of timber must be $10d$ and the maximum fastener diameter d must be 12 mm.

(c) When nailing wood-based sheathing materials to a timber backing to form shear walls or floor diaphragms, the minimum sheathing thickness must be $4d$ and maximum fastener diameter d 3.1 mm.

(d) Some relaxations are permitted on the previous two rules, but lower q values then apply.

Medium- and high-ductility structures must be checked on capacity design principles to ensure that yielding occurs in the connections intended to yield, and that other parts of the structure remain elastic. These elastic parts include the timber members themselves, although EC8 refers particularly to the following

(a) anchor-ties and any connections to massive sub-elements

(b) connections between horizontal diaphragms and lateral load-resisting vertical elements.

The overstrength factor to be used is not stated in EC8; it is merely required to be 'sufficient'. The factor should depend on the possible variation in connection strength; if its as-built strength could exceed the design strength considerably, then an appropriately large overstrength factor is needed. For connections using steel connectors where the strength variation is low, the overstrength factor of 1.25 recommended by EC8 for steel structures is suggested as appropriate.

11.4.2   Practice in the USA

By contrast with the Eurocode, US codes for seismic-resisting timber design are based on extensive application rules rather than broad principles. An extensive summary is provided by Dolan (2003). Dolan lists the following US design standards and industry guidance.

(a) *NEHRP Recommended Provisions for Seismic Regulations for New Buildings and Other Structures* (Building Seismic Safety Council 2000), which forms the basis for the IBC: 2003 (ICC 2003) and other model US codes.

(b) *ASD Manual for Wood Construction* (American Forest and Paper Association 1999).

(c) *Timber Construction Manual* (American Institute of Timber Construction 1994).

(d) *Plywood Design Specification* (APA 1997).

(e) *Engineered Wood Construction Guide* (APA 2001).

The results of an extensive programme of research into the seismic performance of wood-framed buildings, conducted in Californian universities, has recently been published, and contains design recommendations (Cobeen *et al.* 2004).

11.4.3   Bamboo construction

Bamboo (strictly a grass, not a timber) has long been part of traditional construction in seismic regions of South and Central America, where *bahareque* construction consisting of a timber frame supporting clay earth plaster has a mixed, but generally quite favourable record of resisting earthquakes. More recently, more reliable forms

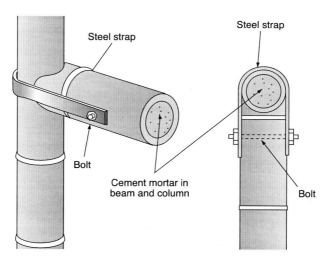

Fig. 11.1   Bamboo beam-to-column connection (after Manual de Construcción Sismo Resistente de Viviendas en Bahareque Encementado)

of bamboo frame have been developed for earthquake-resistant construction. The research has centred on methods of selecting bamboo of sufficient age and therefore strength for construction, cost-effective ways of preventing fungal and insect attack, and methods of connection for bamboo members. The connection methods have been based on the use of steel bolts, with the connection area strengthened by filling the central hollow of the bamboo with a cement mortar (Fig. 11.1). Cement-based renders have been used to enhance the in-plane shear strength of the walls. The Columbian Association of Earthquake Engineering, in conjunction with a number of other organisations, has published a well-illustrated design manual (in Spanish) which is freely downloadable from www.desenredando.org/public/libros/2001/csrvbe/guadua_lared.pdf. The UK-based organisation TRADA International, in partnership with the Indian Plywood Industries Research and Training Institute, has also developed a bamboo building system for earthquake resistance (Jayanetti and Follett 2004).

11.5 References

American Forest and Paper Association (1999). *Allowable Stress Design (ASD) Manual for Wood Construction*. American Forest and Paper Association, Washington DC.

American Institute of Timber Construction (1994). *Timber Construction Manual*, 4th edn. Wiley, New York.

Anagnos T., Rojahn C. & Kiremidjian A. (1995). *NCEER-ATC Study on Fragility of Buildings*. Technical Report NCEER-95-0003. National Center for Earthquake Engineering Research, State University of New York, Buffalo, NY.

APA (1997). *Plywood Design Specification and Supplements*. APA, Tacoma, WA.

APA (2001). *Engineered Wood Construction Guide*. APA, Tacoma, WA.

Building Seismic Safety Council (2000). *FEMA 368 (Provisions) and FEMA 369 (Commentary). NEHRP Recommended Provisions for Seismic Regulations for New Buildings and Other Structures*. BSSC, Washington DC.

CEN (2004a). EN1998-1: 2004. *Design of structures for earthquake resistance*. Part 1: General rules, seismic actions and rules for buildings. European Committee for Standardisation, Brussels.

CEN (2004b). Eurocode 5. EN1995-1-1. *Design of timber structures. General rules*. European Committee for Standardisation, Brussels.

Cobeen K., Russell J. & Dolan J. D. (2004). *Recommendations for Earthquake Resistance in the Design and Construction of Woodframe Buildings*. Consortium of Universities for Research in Earthquake Engineering (CUREE), Richmond, CA.

Dolan J. D. (2003). Wood structures. In Chen W.-F. & Scawthorn C. (eds) *Earthquake Engineering Manual*. CRC Press, Florida.

ICC (2003). *International Building Code* (IBC: 2003). International Code Council, Falls Church, VA.

Jayanetti L. & Follett P. (2004). Earthquake-proof house shakes bamboo world. *Proceedings of the Institution of Civil Engineers, Civil Engineering*, Vol. 157, No. 3, August, p. 102.

Karacabeyli E. & Popovski M. (2003). Design for earthquake engineering. In: Thelandersson S. & Larsen H. J. (eds) *Timber Engineering*. Wiley, Chichester.

Scawthorn C. *et al.* (2005). *Preliminary Observations on the Niigata Ken Chuetsu, Japan Earthquake of October 23, 2004*. Earthquake Engineering Research Institute, Oakland, CA. Available at: www.eeri.org/lfe/pdf/japan_niigata_eeri_preliminary_report.pdf.

12 Building contents and cladding

'The damage and/or loss potential of [building contents] so far has not received enough attention, particularly bearing in mind escalating values and value concentrations.'

Herberty Tiedemann. In: *Earthquakes and volcanic eruptions – a handbook on risk assessment.* Swiss Reinsurance Company, Zurich, 1992

This chapter covers the following topics.

- The lessons from earthquake damage
- Analysis, testing and experience databases
- Electrical and mechanical equipment
- Vertical and horizontal services
- Cladding elements

12.1 Introduction

Mechanical equipment, windows, ceilings and cladding may typically represent 70% of a building's value, and its contents can represent many times the value of the building. Failure of these non-structural elements in earthquakes has given rise to financial loss, interruption of business and the loss of essential post-earthquake services. They have also posed a risk to life, either directly due to injury and suffocation from the collapse of false ceilings, cladding elements, etc. or indirectly due to blocking of escape routes. Damage surveys of earthquakes have shown that, in many cases, buildings which have only suffered minor structural damage have been rendered uninhabitable and hazardous to life owing to the failures of mechanical and electrical systems, and damage to architectural elements. For all these reasons, the preservation of non-structural elements may be equal in importance to maintaining the integrity of the building structure.

In some buildings, for example hospitals and other emergency facilities, it is essential that there is no loss of function after an earthquake. In such cases, higher standards of design and analysis will be required both for non-structural elements within the building and also to ensure that essential services such as electricity, water and telecommunications can be maintained. To achieve this, standby generators, fire-fighting water tanks and so on may be required on site as a backup in case the external supplies fail.

12.2 Analysis and design of non-structural elements for seismic resistance

12.2.1 General principles of design and detailing

Non-structural elements may be damaged during an earthquake due to two distinct mechanisms, namely relative displacement or acceleration. 'Displacement-sensitive' elements become damaged by distortions imposed on them by the structure; cladding elements attached to the façade are an example (Fig. 12.1). There are two design strategies that can be employed here. The first option is to make the structure so stiff that the imposed displacements are sufficiently small not to cause damage; limits on building storey drifts are intended to help achieve this. The second option is to make items sufficiently flexible to accommodate the imposed deflections, either by flexibility within the items themselves, or at their points of attachment to the structure.

'Acceleration-sensitive' elements are more compact items for which relative movements between the points of support to the structure are likely to be small,

Fig. 12.1 Failure of a 'displacement-sensitive' non-structural element: façade cladding Mexico City, 1985

but which become damaged due to the accelerations (and hence inertia forces) imposed on them by the structure. Usually, the damage takes the form of the item becoming detached from its support. The design strategy is then to make the anchorage of the items strong enough to develop the shear and overturning forces needed to prevent failure.

Design may proceed by means of an analysis of the displacements and/or accelerations that the non-structural element has to accommodate. For essential plant items, such as standby generators in critical facilities, this may need to be supplemented by direct testing on a shaking table, or by reference to databases recording experience of plant performance in earthquakes. These types of approach are discussed in the following sections.

For standard items, a more qualitative approach is often the most valuable; much damage to items such as false ceilings, storage shelves and cabinets can be avoided by the use of inexpensive holding-down bolts and restraints. The checklists for non-structural items given by ASCE/SEI 31-03 (ASCE 2003) are a useful source of qualitative design and assessment information.

The use of seismic isolation to protect building contents is covered in Chapter 13.

12.2.2 Analysis for displacement-sensitive items

In principle, the requirement is to establish the maximum relative displacement between points of attachment. When the deformations are derived from a response spectrum analysis, using the Eurocode 8 (CEN 2004) design spectrum, the design relative displacement d_r is given by equation (12.1); the meaning of the notation symbols on the right-hand side of the equation is explained in the next paragraph.

$$d_r = q\nu\sqrt{(d_{r1}^2 + d_{r2}^2 + d_{r3}^2 \dots)} \tag{12.1}$$

In equation (12.1), the relative displacement d_r is calculated from the SRSS combination (or, if necessary, CQC combination – see subsection 3.2.6) of contributions to the relative displacement in each mode, d_{r1}, d_{r2}, d_{r3}, etc., calculated from a response spectrum analysis. Note that it is usually unconservative to calculate the relative displacement as the difference between maximum displacement calculated between the points of interest (e.g. the top and bottom of a storey). Such a practice would underestimate the contribution of higher modes of vibration to relative displacement, which may be significant at the upper levels of tall buildings. The calculated deformation from the response spectrum analysis must be multiplied by the behaviour factor q, to allow for the post-yield deformation of the structure. The modal contributions d_{r1}, d_{r2}, d_{r3} are calculated for the ultimate limit state (ULS) event, which usually has a 475-year return period. In Eurocode 8, however, the design displacement is reduced by a factor ν. This is because limiting damage to non-structural elements is usually a serviceability limit state (SLS), rather than a ULS consideration, and a shorter return period is appropriate. The recommended reduction factor in Eurocode 8 is generally 0.4, but rises to 0.5 for important buildings such as hospitals.

In practice, protection to many standard architectural elements such as cladding and partitions is provided by meeting code-specified limits on storey drift (the relative displacement between the top and bottom of a storey). In Eurocode 8 Part 1, where cladding elements are rigidly attached to the structure, the SLS storey drift is limited to 0.5% of storey height but this rises to 0.75% for rigidly attached ductile cladding. Where the cladding fixings can accommodate the structural deformations, the drift limit rises to 1%.

Extended items such as pipes which have multiple supports to the structure may require more sophisticated analysis; this is briefly discussed in section 12.4.

12.2.3 Analysis of simple acceleration-sensitive items

During an earthquake, an item of equipment inside a building, such as a pump, will generally experience different motions to those of a similar piece of equipment attached to the ground outside. As the building sways in the earthquake, the accelerations at ground-floor level will generally be similar to those of the ground outside, but will change up the height of the building, generally becoming greater, except in tall, flexible buildings or those with base isolation. Not only is the amplitude of motion affected, but so too is its frequency content.

Codes of practice give simple formulae allowing for the variation in amplitude of motion with floor level and the modification in frequency content. These formulae are adequate for most practical cases found in standard buildings, for the purposes of designing floor fixing for simple items such as pumps or cabinets. They can also be used for the inertia term for cladding fixings; however, the fixings will experience additional forces due to imposed deformation.

The equation in Eurocode 8 is as follows

$$F_a = (S_a W_a \gamma_a)/q_a \tag{12.2}$$

where F_a is the horizontal seismic force, acting at the centre of mass of the non-structural element in the most unfavourable direction; W_a is the weight of the element; γ_a is the importance factor of the element, equal to 1 for most items, but rising to 1.5 for critical items; q_a is the behaviour factor of the element, which varies between 1 and 2; and S_a is the 'seismic coefficient', which allows for the difference between the peak acceleration on the ground outside the building, and that experienced by the non-structural element. S_a is calculated as follows

$$S_a = \alpha S[3(1 + z/H)/(1 + (1 - T_a/T_1)^2) - 0.5] \tag{12.3a}$$

or

$$\alpha S \text{ if greater} \tag{12.3b}$$

where α is the ratio of the design ground acceleration on hard ground a_g to the acceleration of gravity g for ULS; S is the soil factor, which ranges from 1 for hard ground to 1.8 on soft ground; T_a is the fundamental vibration period of the non-structural element; T_1 is the fundamental vibration period of the building in the relevant direction; z is the height of the non-structural element above the level of application of the seismic action; and H is the building height measured from the foundation or from the top of a rigid basement.

For a rigid item such as a cabinet or plant item fixed rigidly to the structure without anti-vibration mounts ($T_a = 0$), it is easy to show that S_a equals αS at ground level and $2.5\alpha S$ at the top of the building. In fact, this is a simplification; the true amplification at the top of the building will in fact depend on the damping in the building, how close the building is to resonance with the earthquake motions and the degree of yielding within the structure. In most cases, however, an amplification of 2.5 at the top of the building is reasonably conservative.

Where the item has significant flexibility – a pump on anti-vibration mounts, for example – the amplification at the top may increase above 2.5, and according to the Eurocode 8 formula reaches 5.5 where there is a perfect match between building period and non-structural item period ($T_a = T_1$). Again, this is an approximation, but sufficient for most cases.

12.2.4 Analysis of acceleration-sensitive items using 'floor response spectra'

The Eurocode 8 formulae given as equation (12.2) and (12.3) above (and similar ones in other seismic codes) assume that the non-structural element being considered is simple enough to be approximated by a single degree of freedom system. More complex cases may need more sophisticated methods of justification, if they are critical. Qualification by testing and using 'experience databases' are two such methods discussed in the next two sections. These are methods that dispense with the need for sophisticated analysis. However, analysis routes to qualification are available. One obvious method would be to include the non-structural item directly in the model for the main structural analysis. There are difficulties here, however. One problem might be in ensuring that sufficient modes of vibration have been considered in the analysis to capture adequately the response of one relatively small part – that is, the non-structural item. A more practical (and insuperable) objection to this route is that the details of the non-structural element may well not have been finalised at the time of the structural analysis. The solution here is to use the structural analysis to produce 'floor response spectra' at the points where the non-structural items are expected to be attached. These floor response spectra are produced exactly in the same way as normal ground spectra, but they relate to the motions of the structure at the attachment points. Response spectra express both the amplitude and frequency content of motions, and so include the factors allowed for in a more simple way by code formulae such as equations (12.2) and (12.3).

The most direct way to derive floor response spectra is to carry out a time-history analysis on the main structure, which will then yield the time history of motions at the attachment points. This, however, requires at least one input time history at ground level. Usually, an input response spectrum will be specified and the appropriate choice of suitable time histories compatible with the design spectrum is not straightforward, as discussed in subsection 2.8.3. As an alternative, ASCE 4-98 (ASCE 1998) refers to a number of methods for producing a floor response spectrum directly from a ground spectrum and a structural model, which avoids having to select a time history. One such direct spectrum-to-spectrum method is given by Singh (1984).

The floor response spectra generated by either time history or direct spectrum-to-spectrum methods will show a strong peak at the first mode period of the building structure. This period, however, is likely to be subject to considerable uncertainty, and a broadening of the peak to allow for this uncertainty is advisable; the reader should refer to the relevant rules in ASCE 4-98 (ASCE 1998).

Floor response spectra are an appropriate tool where the building response is expected to remain linear during the design earthquake, and they are often used for applications such as nuclear power plants where this is a requirement. Where substantial non-linear response is expected, the direct spectrum-to-spectrum method of producing floor spectra is unlikely to be appropriate, and the time-history analysis would need to be non-linear.

It can be seen then that the use of floor response spectra is a way of decoupling the analysis of the main structure from that of the non-structural element, which is particularly useful when details of the former must proceed before those of the latter are available. The method implicitly assumes that the response of the non-structural item will not significantly affect that of the main structure. Usually, an estimate of the mass of the item can be made with sufficient accuracy at the time of the main structural analysis, and the implicit assumption is valid. However, where natural periods of the structure and non-structural item are very close, the assumption breaks down, even if the mass of the non-structural item is an order of magnitude less than that of the structure. The only way to treat this case of the 'tail wagging the dog' is to abandon floor response spectra, and analyse the system as a coupled whole.

12.2.5   Testing of acceleration-sensitive items

The analysis methods discussed in the two preceding sections are appropriate for designing fixings and structural members but they are most unlikely to give an insight into whether the mechanical parts of a plant item will continue to function during and after an earthquake. It is easy (and inexpensive) to design the holding-down bolts for an emergency generator set which will ensure that it remains fixed to the structure, but will it still produce electricity?

In fact, rotating machinery is robust, and generally continues to function if it does not become detached. A ship in a storm can experience accelerations of the same order as those in an earthquake, and yet its machinery generally continues to function, and hard drives of computers also continue to spin. However, some plant may be more sensitive; for example, mechanical relays found in older electrical switchgear may malfunction, and some electrical insulation is brittle and prone to fracture. The most direct way to ensure that a plant item performs adequately is by placing it on an earthquake shaking table, and subjecting it to suitable motions. For plant items within a building, these motions would need to correspond to the floor response spectra discussed in the previous section. This method of direct qualification has been used for safety-critical plant items in nuclear power plants.

12.2.6   Qualifying acceleration-sensitive items from 'experience databases'

Testing of plant on a shaking table is expensive, and in any case most plant items are fairly robust against seismic motions. During the 1980s, the nuclear power industry in the USA started to develop a database which showed how plant items had fared in previous earthquakes. The idea was to qualify a standard piece of equipment in a nuclear power plant by showing that the similar kit in conventional power stations had survived earthquakes which bounded the design motions. Of course, some plant in a nuclear station is highly specialised but others, such as pumps and standby generators, are standard items which may still be required to perform a vital safety-related function needing to be preserved during and after an earthquake.

It was for such items that the Seismic Qualification Utilities Group (SQUG) produced the 'generic implementation procedure (GIP) for seismic verification of nuclear power plant' (Starck and Thomas 1990). The procedure allows a plant item to be qualified from the database assembled, if the following conditions are met.

(a) The design spectrum is bounded by the estimated survival spectrum in the database.
(b) The plant item is similar in design to one of the items listed in the database.
(c) The plant item is adequate in terms of security and rigidity of fixing, and workmanship.
(d) The plant item does not interact adversely with other items. It must not be damaged by knocking against adjacent items, or by debris falling on to it.

Unfortunately, the SQUG database is not in public circulation, and is only available to subscribing members, for which a substantial subscription is necessary. However, a useful checklist in the public domain for non-structural items is given by ASCE/SEI 31-03 (ASCE 2003), referred to earlier; this is mainly based on observations of performance during past earthquakes.

12.3 Electrical, mechanical and other equipment

Plant items such as generators, pumps and computers have been referred to in previous sections; usually it is sufficient to ensure that they are adequately fixed to the structure. Small tanks holding liquid may be treated in the same way. However, larger tanks, particularly if they hold flammable liquids, need to be designed allowing for the interaction between the tank and its contents, which is complex. A New Zealand report (Priestley et al. 1986) published some years ago still forms one of the best sources of advice on tank design, and there is also information in Eurocode 8 Part 4.

12.4 Vertical and horizontal services

Extended non-structural elements within a building, such as pipes, ducts and lifts are attached to many points of the structure, and the simple type of analysis described in subsection 12.2.2 is not applicable because the imposed deformations are more complex. Small pipes are likely to be sufficiently flexible to accommodate these imposed deformations without distress, although where buried services enter a building, it may be necessary to introduce additional flexibility. Explicit analysis of multiply-supported systems is given in ASCE 4-98 (ASCE 1998).

Automatic shutdown valves are available at moderate cost intended to shut off gas supplies if the ground acceleration exceeds a threshold such as 10% g. These may be useful to reduce the risk of fire and explosion in the building.

Once again, useful design information is given in ASCE/SEI 31-03 (ASCE 2003).

12.5 Cladding

Failure of the external cladding to buildings occurs frequently during earthquakes. It causes a serious falling hazard to people outside the building and may render the building itself temporarily uninhabitable.

Cladding elements are vulnerable to both imposed deformations in their plane and to accelerations normal to their plane. Cladding attached to the outside of the structural frame in general needs to be checked for both effects. Externally attached precast concrete panels may be particularly vulnerable; their in-plane rigidity can give rise to high deformation forces, while their large mass causes high out-of-plane inertia forces. One solution is to provide slotted connections at the base of the panels which allow in-plane movement, releasing the associated forces, but providing out-of-plane restraint. For glass curtain walling in a secondary metal restraint frame, special seismic gaskets have been developed which allow the glass to move relative to the restraint frame. External stone or brick cladding needs to be well tied back to the main structure for out-of-plane inertial restraint, while providing sufficient structural stiffness to limit storey drifts to the appropriate code-specified limits (subsection 12.2.2) should limit damage due to imposed in-plane deformations.

Concrete or steel frames with infill masonry forming the cladding are discussed in subsection 8.7.8.

12.6 References

ASCE (1998). ASCE 4-98. *American Society of Civil Engineers Standard: Seismic analysis for safety-related nuclear structures*. ASCE, Reston, VA.

ASCE (2003). ASCE/SEI 31-03. *Seismic evaluation of existing buildings*. American Society of Civil Engineers, Reston, VA.

CEN (2004). EN1998-1: 2004. *Design of structures for earthquake resistance*. Part 1: General rules, seismic actions and rules for buildings. European Committee for Standardisation, Brussels.

Priestley M. J. N. *et al.* (1986). *Seismic Design of Storage Tanks*. New Zealand National Society for Earthquake Engineering, Wellington.

Singh M. P. (1984). State-of-the-art of development of floor spectra. *Proceedings of Structural Engineering in Nuclear Facilities Conference*, sponsored by the Committee on Nuclear Structures and Materials of the Structural Division of ASCE, Raleigh, NC, 10–12 September. Ucciferro J. J. (ed.), ASCE, New York, Vol. 1, pp. 478–494.

Starck R. G. & Thomas G. G. (1990). Overview of the SQUG generic implementation procedure (GIP). *Nuclear Engineering and Design*, Vol. 123, Issues 2–3, October, pp. 225–231. Elsevier Science Publishers, North Holland.

13 Seismic isolation

'Seismic isolation is a design strategy based on the premise
that it is both possible and feasible to uncouple a structure
from the ground, and thereby protect it from the damaging
effects of earthquake motions.'

Ian Buckle and Ronald Mayes. In: Seismic isolation:
from idea to reality. *Earthquake Spectra*, 1990,
Vol. 6, No. 2, May, p. 6

This chapter covers the following topics.

- Principles of seismic isolation
- Lessons from 30 years of seismic isolation
- Types of seismic isolation and their application
- Analysis of base-isolated structures
- Standards for testing and design
- Active and semi-active control

13.1 Introduction

13.1.1 Seismic isolation – an idea whose time has come

The idea that a building could be protected from earthquakes by decoupling it
from the ground with an isolation layer dates back to Roman times; Pliny the
Elder wrote in his famous *Natural History*: 'The temple of Diana in Ephesus has
been built on a marshy soil to protect it from earthquake and fault effects. Between
the soil and the foundations of the temple a layer of coal and wool fleeces has been
interposed.' (Pliny was later killed while observing the eruption of Vesuvius in
AD 79; the account he wrote of the eruption luckily survived him and has proved
invaluable to succeeding generations.) In the nineteenth century, John Milne, an
English pioneer of engineering seismology working in Tokyo, constructed a
building where the isolation layer consisted of four cast iron balls, although it
has to be admitted he later declared the attempt a 'failure' (Muir Wood 1988).
However, it was not until the development and use of elastomeric lead–rubber
bearings in New Zealand in the mid 1970s that seismic isolation became a practical
reality; since then, according to Mayes and Naeim (2001), it has demonstrated all
the characteristics of an idea whose time has come, with hundreds of buildings and
bridges in many seismic areas of the world employing the principle. The evidence

from both testing laboratories and real earthquakes (the latter still rather limited) suggests that the principle lives up to its promise.

13.1.2 Basic principles of seismic isolation

The way in which seismic isolation works is, in concept, straightforward (Fig. 13.1). In the same way that shock absorbers smooth out the ride of a car by absorbing the bumps in a rough road, seismic isolation works by decoupling a building from the violent ground motions caused by an earthquake. Perfect isolation – for example, a building on a frictionless surface – would leave the building completely unaffected by the earthquake, but would not of course be practical; for one thing, the building would then not be able to resist lateral forces such as wind loads. However, very substantial reductions are possible by mounting a building on bearings with high but finite horizontal flexibility. The bearings act to lengthen the natural

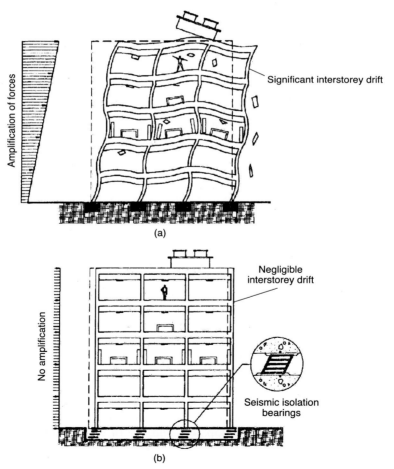

Fig. 13.1 Basic principles of seismic isolation: (a) conventional building; and (b) isolated building (from Mayes and Naeim 2001)

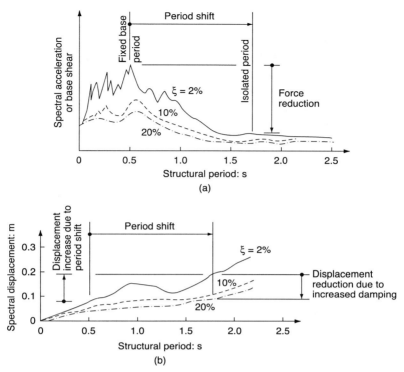

Fig. 13.2 Effect of period lengthening and increased damping on response for earthquake motions with a predominant period of around 0.5 s: (a) effect of period shift on design forces; and (b) effect of period shift and damping on relative displacement between ground and structure

period of the building, taking it away from the main periods of the ground motion (Fig. 13.2). Usually, the isolation plane is at the base of the building – hence the common term 'base isolation' – but even in buildings this is not always the solution adopted, and in bridges the isolation plane usually occurs near the top, at the deck support points.

A major advantage of seismic isolation is that it protects all the elements above the isolation plane. The reduction in accelerations above the isolation plane not only reduces the inertia forces that the structure must resist, but it also reduces the forces on attachments such as water tanks or plant items, and so these too are less prone to failure. The reduction in structural forces also reduces the shear deformations in the structure, and hence the damage to cladding, glazing, partitions and other non-structural elements. Thus, seismic isolation serves to protect both the building's structure and its contents, and the building is much more likely to be able to function normally immediately after a strong earthquake.

Relative deflections within the superstructure are reduced at the expense of large deformations occurring across the isolation plane, and these must be considered in design. Services (e.g. water and gas pipes) crossing into the building must be flexible enough to accommodate these deformations. An 'isolation gap' must be created and maintained between the building and the ground. It is also essential to ensure that the deformations do not become so large that they compromise the ability of the isolation bearings to carry the building's weight. In order to limit deformations across the isolation plane occurring during the design earthquake to a few hundred millimetres (a typical value for bearing capability), additional damping within the bearing is often provided (Fig. 13.2). It is also desirable that the bearings should return to their original position after an earthquake. Practical means of achieving these objectives are discussed later.

So far, the discussion has been in terms of reducing the horizontal effects of earthquakes. What about the vertical motions to which all earthquakes give rise? In practice, it is much less important to provide protection against these vertical effects. All buildings, by necessity, are built with a substantial factor of safety against gravity loads – which are equivalent to 100% g vertically – but usually only need to resist horizontal wind forces equal to a few per cent of their weight. Moreover, buildings are much stiffer vertically than horizontally, and vertical deformations in an earthquake are unlikely to distress cladding and other non-structural elements. Therefore, protection against vertical seismic motions is not usually needed. In fact, it is desirable to make seismic isolation bearings very rigid in a vertical direction, because vertical flexibility in the bearings would cause the building to rock during an earthquake, negating some of the benefits of horizontal isolation.

In summary, seismic isolation protects both the building and its contents by detuning it away from the main forcing periods of earthquake ground motions. The bearings usually need to be provided with additional sources of damping, to limit deflections across the isolation plane to safe proportions. They should preferably be designed to re-centre after an earthquake, and must be able to resist normal lateral forces due to wind without causing damage. The seismic deformations across the isolation plane need to be considered in the design of services crossing into the building, and an isolation gap needs to be provided and maintained around the building.

13.1.3   Applications in practice

Seismic isolation has been widely used in new buildings in New Zealand, USA, Japan, Indonesia, Italy (Fig. 13.3) and elsewhere. However, the current perception is that additional costs of isolation mean that the total construction cost is larger than for conventional fixed-based buildings; Mayes and Naeim (2001) suggest a premium of up to 5% of structural cost is typical compared to a fixed-base building. The cost of the bearings in relation to the total building cost (including contents and land) will however be less, and the extra cost of the bearings may be offset, at least to some extent, by savings in the foundations and from not having to provide seismic detailing in the superstructure.

Mayes and Naeim (2001) also point out that the performance of an isolated building will be superior to its fixed-base equivalent. In fact, most new base-isolated building projects have been for special or major buildings where the ability to function immediately after an earthquake is particularly important. If the technology becomes more familiar, and perhaps if design standards become less

(a)

(b)

Fig. 13.3 Telecom Administration Centre, Ancona, Italy: (a) external view; and (b) interior detail showing bearing

Fig. 13.4   Retrofitting of an existing reinforced concrete building. Multifunction centre, Napoli, Italy

onerous relative to those for fixed-base buildings, it may be that seismic isolation will become much more widely used for standard projects, but this has yet to happen.

Seismic isolation has also been widely used to protect existing buildings with inadequate seismic resistance (Fig. 13.4). The attraction is that structural intervention is concentrated at the isolation plane, and the need to strengthen elements elsewhere is much reduced or even eliminated. This concentration of effort reduces the disruption during the retrofitting works (a great advantage if the building has to remain in operation during this period) and also reduces the architectural impact, which is likely to be of crucial importance in historic buildings where the original features must be preserved. Individual isolation of precious items in museums has been adopted in California and elsewhere.

Seismic protection of bridge decks has also been widely used by introducing an isolation plane at the tops of piers and abutments. Since most bridges require some sort of bearing at this point to allow thermal movements, the technology is particularly appropriate, and in consequence is a more standard procedure than for buildings.

13.2   Lessons from 30 years of seismic isolation

13.2.1   Introduction

Since publication of the first edition of this book in 1988, much has changed. Many more structures in seismic areas have been built with seismic isolation, a number of standards governing their design have been published and there is limited experience of how isolated buildings actually fare in strong earthquakes.

13.2.2   Performance of seismic-isolated buildings in earthquakes

Five Californian buildings experienced moderate to high ground motions during the 1994 Northridge earthquake (Smith 1996). Three performed very well, most notably the University of Southern California (USC) Hospital, where the peak ground acceleration (pga) near the site was recorded at 0.49 g. The eight-storey building was mounted on lead–rubber and elastomeric bearings, which increased its first natural period from 0.8 to 2 s. The hospital was functioning immediately after the earthquake, with reports of only minimal damage. Less successful were two, three-storey steel-framed houses, which were mounted on helical steel springs with viscous dampers. It appears that there was non-structural damage, although the structures survived during motions which were probably greater than those at the USC Hospital. The rocking introduced by the vertical flexibility of the springs has been suggested as a cause for the damage.

Two Japanese buildings were in the epicentral area of the 1995 Kobe earthquake and both performed well. One was the West Japan Postal Computer Centre, a six-storey, 47 000 m^2 building, mounted on 120 elastomeric bearings with steel and lead dampers, which gave an isolated period of 3.9 s. The building was undamaged by the earthquake ground motions, recorded with a pga of 0.4 g at the site, while a neighbouring fixed-base building reported some damage.

13.2.3   The regulatory position

Eurocode 8 Part 1 (CEN 2004) provides rules for the design of new seismically isolated buildings, and Part 2 does the same for bridges. A new European Product Standard, EN 15129: 200X (Anti-seismic devices), will cover the testing and specification of anti-seismic devices, including isolation bearings; testing is addressed by an annex in Eurocode 8 Part 2.

The US Standard ASCE 7-02 (ASCE 2002) gives rules for the design of seismically isolated structures, which are adopted for buildings by the International Building Code (ICC 2003). ASCE Standard ASCE 7-02 provides testing requirements for seismic bearings in buildings. For bridges, there is an AASHTO specification for design (AASHTO 2000), and the Highways Innovative Technology Evaluation Center (HITEC 1996) has produced guidelines for testing, to which the AASHTO specification refers.

13.3   Seismic isolation systems

13.3.1   Functional requirements of a seismic isolation system

An isolation system needs to provide the following.

(a) Horizontal flexibility to lengthen the building period, while maintaining
    vertical stiffness.
(b) Damping, to restrict the relative deformation at the plane of isolation and
    limit it to within the capacity of the bearings.
(c) Sufficient stiffness to prevent damage under wind forces.
(d) It is also desirable that residual horizontal deflections of the building relative
    to the ground are small after an earthquake.

These four aspects are now discussed in turn.

13.3.2 Providing horizontal flexibility

Horizontal flexibility can be provided by rubber or sliding bearings, as described in
more detail in the next section. Less conventionally, slender structures can be
allowed to lift off their bearings. This system has been used for a bridge (Beck
and Skinner 1974) and a chimney (Sharpe and Skinner 1983) in New Zealand
and more recently for a building in Tokyo (Fig. 13.5; Lang *et al.* 2001). Slender
bearing piles in oversized casings have also been used, most notably in Union
House (Fig. 13.6), a 12-storey office building in Auckland (Boardman *et al.* 1983).

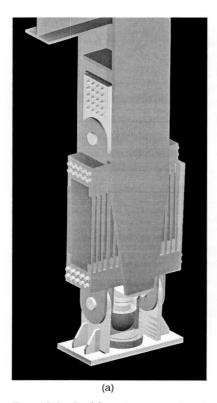

(a)   (b)

*Fig. 13.5 Building incorporating 'stepping' columns, La Maison Hermes, Tokyo:
(a) schematic; and (b) column during installation (see Lang et al. 2001)*

Fig. 13.6 Steel hysteretic damper at pilehead, Union House, Auckland

13.3.3 Providing damping and initial stiffness

Various means of providing damping exist, which can be classified as hysteretic, Coulomb (frictional) or viscous. Hysteretic dampers consist of ductile metal elements designed to yield, and the lead–rubber bearings described below fall into this category. Dampers based on yielding of steel have also been quite widely used (Fig. 13.6). The friction in sliding bearings provides Coulomb damping. The advantage of both hysteretic and Coulomb damping is that they provide the high initial stiffness (that is, before the metal yields, or the friction is overcome) which is needed for wind resistance. These two forms of damping are independent of velocity, at least to a first order; by contrast, viscous damping is zero without a relative velocity across the damper, but increases as that velocity becomes larger. A resulting advantage of viscous damping is that it is at maximum when the deformation and acceleration are at their minimum. Hence the damping and inertial forces are out of phase, and so the structure does not have to resist the maximum of the two effects simultaneously. Viscous dampers can take the form of conventional fluid-filled dampers (similar in principle to a car's shock absorber). The properties of elastomeric rubbers can also be modified to produce damping ratios of up to 20% of critical. Elastomer properties can also be modified to produce sufficient initial stiffness for wind loading, while softening considerably at larger deformation to provide the necessary period lengthening under severe earthquake loading.

13.3.4 Providing re-centring

Rubber bearings with viscous damping will always tend to return to their initial position after an earthquake. With hysteretic dampers, the bearings must have enough horizontal stiffness to recover sufficient of the plastic deformation in the dampers. The residual deformation when the system comes to rest cannot exceed the deformation of the bearing under the action of the yield force in the damper, and so the horizontal stiffness of the system must be sufficient to keep this within acceptable limits. Similar considerations apply to systems with frictional damping. In planar sliding bearings, there is no tendency for a re-centring force to develop and supplementary elastic systems may be added to provide this, for example by means of elastomeric bearings combined in series with the sliding bearings. Alternatively, the sliding bearings may have a spherical shape, as described below for the friction pendulum system (Fig. 13.9), which again provides a re-centring force.

Requirements for re-centring in bridges are given in the forthcoming Eurocode 8 Part 2, although there is no equivalent requirement for buildings in Part 1. (Note however that the consequences of a bridge deck falling off its seismic isolation bearings are likely to be much more catastrophic than would be the case for a building.) The re-centring rule for bridges in EC8 Part 2 requires that when the isolation system is pushed to its maximum deflection capacity d_m and then released, the residual deflection d_{rm} should be such that more than half of the design deflection under seismic loading is recovered. (Note that the design deflection is likely to be less than the deflection capacity d_m and of course must not exceed it.) The rule also requires that the increase in resistance of the system as its deflection increases from $0.5d_m$ to d_m should exceed 1.5% of W_d (d_{rm}/d_m), where W_d is the weight of the superstructure mass. This is to ensure that the system always has a small positive stiffness tending to return it to its centre. Note that individual components of the isolation system do not have to obey these rules, as long as the combined effect of all the components ensures re-centring. For example, simple pot bearings (which have no ability to re-centre) can be combined with rubber bearings, which provide the re-centring.

13.3.5   Types of seismic isolation bearing

Figure 13.7(a) shows a lead–rubber bearing schematically. The layers of rubber provide the lateral flexibility, while the steel plates restrain the rubber from bulging outwards under vertical loading and so help maintain vertical stiffness. The lead plug provides hysteretic damping after it has yielded, and high initial stiffness before yield. The plug is stopped short of the top of the bearing to prevent it carrying vertical load. The base plate is fixed to the substructure and the top plate is usually dowelled to the superstructure in such a way as to achieve the necessary horizontal shear transfer, but to prevent uplift (vertical tension) forces developing. A horizontal deflection capacity of around 1 to 1.5 times the net rubber thickness (i.e. after deducting the thickness of the steel plates) can be achieved in lead–rubber bearings, and an equivalent viscous damping ratio of up to around 30% of critical can be obtained.

High damping rubber bearings (Fig. 13.7(b)) are constructed in a similar way, but have no central lead plug. The elastomer forming the bearing is made with filler materials which modify the damping and also stiffness, providing high initial stiffness but lower horizontal stiffness at seismic deflections (Fig. 13.8). The viscous damping ratio achievable, at around 20%, is less than for lead–rubber bearings, but the deflection capacity is greater at up to twice the net rubber thickness.

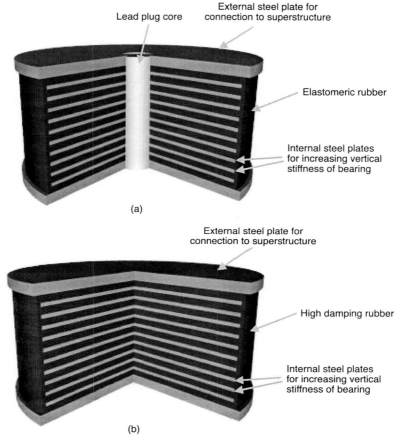

Fig. 13.7 Rubber seismic isolation bearings: (a) lead–rubber bearing; and (b) high damping rubber bearing

This larger deflection capacity arises not for reasons of overall stability, but from the lower heat dissipated per deflection cycle to a given limit; this leads to a lower temperature rise.

Friction pendulum bearings are a patented system produced by Earthquake Protection Systems Inc. of California. The system consists of an articulated slider with a low-friction coating moving on a spherical stainless steel surface supporting the structure (Fig. 13.9). The bottom of the slider is also housed in a low-friction spherical bearing which allows it to rotate and maintain good contact with the upper plate. The restoring force arises because the spherical nature of the upper bearing plate causes it to rise when it moves relative to the slider, and the weight of the building then causes the system to return to the central position. In fact, the effective spring stiffness can easily be shown to depend on the radius of curvature of the upper plate and the vertical force it supports. When the vertical force arises only from gravity, again it is easy to show that the period of the system T depends only on this radius, just as the period of a pendulum depends only on its

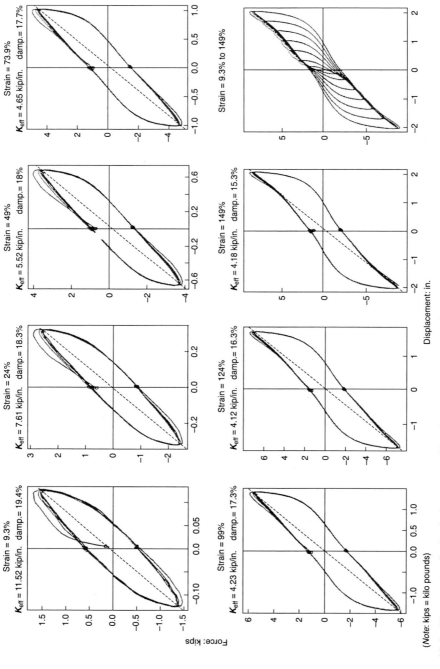

Fig. 13.8 Hysteresis of a high damping rubber bearing at strains of up to 149%

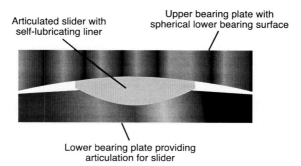

Fig. 13.9   Friction pendulum system

length and not the weight at its end. This property gives rise to the name 'friction pendulum'. T can be shown to be related to the radius R by

$$T = 2\pi\sqrt{R/g} \tag{13.1}$$

In fact, since the restoring and frictional forces are also proportional to the vertical force, an advantage claimed for the system is that the horizontal forces will always act through the centre of mass, thus minimising torsional effects. However, during an earthquake, vertical forces on the bearings may vary considerably due to seismic overturning moments, and so in practice response may be more complex. Also, local areas with high gravity loads will attract high lateral forces, which may not be desirable.

The deflection capacity of friction pendulum systems is limited by their size in plan. The equivalent viscous damping ratio ξ depends on the friction coefficient μ, the deflection and the radius of the bearing R; at maximum deflection D, ξ can be shown to equal approximately

$$\xi = \frac{2}{\pi}\left(\frac{\mu}{4\pi^2 D/(gT^2) + \mu}\right) \tag{13.2}$$

ξ increases with the period T and friction coefficient μ, but if μ is too large, the bearing will tend to 'stick' before returning to its central position. In fact, a displacement of at least μR is needed to ensure that the system moves back towards the centre from rest. Taking into account practical limitations on T and μ, the maximum achievable damping ratio is therefore around 20%. There is likely to be more variation in μ than in the damping value of rubber bearings, resulting in a big variation in effective structural damping ξ, although the period T is precisely controlled through the radius of the bearing.

13.4   Design considerations

13.4.1   Suitability of buildings for base isolation

(a) Height limitations
Figure 13.2 shows that seismic isolation works by providing a large separation between the period of a building and the period of the ground motions to which it is subjected. Therefore tall buildings are not usually suitable for base isolation,

since they already have long periods. Moreover, there could be practical difficulties in designing the bearings for the large vertical loads they are likely to carry. Typical upper limits on height are quoted by Mayes and Naeim (2001) as 8–10 storeys for unbraced frame buildings and 12–15 storeys for shear wall buildings.

(b) Soil conditions
Sites which are subject to particularly long period earthquake motions are also not suitable for isolated structures, and in this case lengthening the building period might even cause an increase in seismic demand. Therefore, buildings on very soft soil sites, which amplify the long period content of the earthquake motions, are more likely to benefit by being designed as stiff as possible.

(c) Wind loads
Where the design wind load exceeds 10% of the building weight, seismic isolation is unlikely to be attractive, since the building needs to respond relatively rigidly to wind loading, and therefore a flexible response would only be possible for earthquake excitations higher than the wind load, which would limit the effectiveness of the isolation. However, design wind loads exceeding 10% of weight are unusual in engineered buildings, particularly concrete ones.

(d) Overall slenderness
Slender buildings may be unsuitable for isolation with conventional isolation bearings, since the bearings would need to be designed for large vertical forces due to rocking, which might not be feasible, both in compression and tension. Other types of isolation system may however work; for example, the stepping system shown in Fig. 13.5 depends on high slenderness.

(e) Building separations and structural joints
An isolated building has substantially larger horizontal deflections than its fixed-base equivalent (Fig. 13.2), and therefore greater separations from adjacent structures are required than is the case for non-isolated buildings. A gap of at least 200 mm (and often more) is needed, and constricted sites where this is not possible may not be suitable. This may also prove an issue for expansion joints within a building; here, high-viscosity dampers (called 'lock-up devices'), are a possible solution. These allow slow relative movements across joints due to thermal and creep effects to take place, but lock up to prevent impact during the much more rapid movements during an earthquake.

(f) Building function
One of the benefits of isolation is that the building contents, as well as its structure, are protected. Therefore, isolation may be particularly advantageous where the building contents are valuable and its ability to function after an earthquake is crucial. Current seismic codes in fact require that isolated building superstructures (unlike fixed-based ones) have an essentially elastic response to earthquakes; this is partly due to conservatism and partly due to the fact that a ductile response is much less effective in an isolated structure than a fixed-based one (see subsection 13.5.4). The seismic performance standard achieved from a code-designed isolated

structure is therefore likely to be substantially higher than for the equivalent fixed-based structure. This makes comparisons between fixed-base and isolated options for a given structure more than usually complex, because life-cycle costs including repair and loss of use are involved and not just initial construction costs.

13.4.2   Suitability of isolation as a strengthening technique

The same considerations of suitability discussed above for new buildings apply when considering isolation, as opposed to member strengthening, as a technique for improving the seismic performance of an existing building. Isolation has the added advantage that operations are confined to one level (the isolation plane) which should reduce disruption to the normal functioning of the building, particularly if the isolation plane can be in the basement. It is also highly suitable for buildings where the main deficiency is a lack of seismic detailing, since the ductility demands in the superstructure are greatly reduced by isolation. Isolation on its own, however, is unlikely to rectify situations where the superstructure has a deficiency in lateral strength of more than 50%; the strength reduction in an isolated building is not as great as Fig. 13.2 suggests at first sight, because the large reduction in spectral acceleration is offset by smaller ductility factors (see subsection 13.5.4). Where there is a large shortfall in lateral strength, additional lateral strengthening measures, for example provision of shear walls, are likely to be required to supplement seismic isolation, which therefore may become relatively less attractive.

13.4.3   Position of isolation level

Often, the isolation level will best be placed below ground level. In this way, none of the above-ground finishes within the building have to cross the isolation plane. Moreover, this position maximises the superstructure protected. Hence seismic isolation is often referred to as *base* isolation. However, other positions for the isolation plane may be indicated, for example when retrofitting existing buildings without a basement. Note that in bridges, the isolation plane is generally between the deck and the pier – that is, at the *top* not the *base* of the structure.

13.5   Analysis of seismic isolation systems

13.5.1   Objectives of analysis

The analysis of a seismic isolation system will usually need to consider two levels of seismic loading: a design earthquake with a return period typically of 475 years and a rare, extreme earthquake with a return period of several thousand years. The objectives of the analysis will primarily consist of the following.

(a) To establish the maximum deformation of the isolator, to make sure that it does not exceed the isolator's limit in an extreme event.

(b) To establish the response of the structure above the isolation plane. Usually, the objective will be to ensure an essentially elastic response in the design event. The response of the structure below the isolation plane must also be considered, usually with the same objective.

(c) To check the performance of the non-structure – finishes, services – particularly where they cross the isolation plane.

13.5.2 Simplified analysis

Since most of the deformation in a seismically isolated building occurs in the isolation bearings, an obvious approximation to its seismic behaviour is to assume that everything above the isolation layer responds rigidly. On this assumption, where the isolation layer is at or near the base of the structure, the building can be modelled as a rigid mass on a spring represented by the bearings. Where the centre of mass coincides with the centre of stiffness of the bearing, the building can be analysed as a single degree of freedom system for each horizontal direction of motion.

The model can be further simplified by representing the isolation system as a linear spring with constant viscous damping. In practice, of course, isolation bearings are often far from linear (Fig. 13.8), but for a first approximation, their combined stiffness can be represented by an equivalent linear spring whose stiffness is chosen to give the total lateral force in the bearings at the maximum deflection under the design earthquake, and a level of viscous damping reflecting the hysteretic energy dissipated (Fig. 3.24). Since the maximum deflection is not at first known, the process is iterative, as follows.

(1) A trial deflection is chosen.
(2) The stiffness and damping of the equivalent linear spring is calculated (Fig. 3.24).
(3) The effective isolated period T_{eff} of the building is calculated as

$$T_{eff} = 2\pi\sqrt{M/\boldsymbol{K}_{eff}}$$

where M is the total mass of the building above the isolation plane and $\boldsymbol{K}_{eff}$ is the equivalent spring stiffness of the isolation system at the trial deflection.
(4) The deflection of the isolation system, ignoring any torsional response can then be calculated as

$$d = \eta S_e(T)M/\boldsymbol{K}_{eff}$$

where $S_e(T)$ is the 5% damped elastic spectral acceleration for the design earthquake at period T. This will usually be taken from the governing seismic code. The implicit assumption is that the foundations are sufficiently rigid in both translation and rotation to transmit the ground motions essentially unmodified to the isolation system; this is an assumption that may need to be checked and if necessary refined at a later stage. Note also that $S_e(T)$ is based on the unmodified ground motion, calculated without structural reduction factors (for example q in Eurocode 8 or R in IBC).

The notation η represents a correction factor to allow for the effect of damping in the isolation system other than 5%. Eurocode 8 gives

$$\eta = \sqrt{10/(5 + \xi)} \geq 0.55$$

where the damping ratio ξ is expressed as a percentage (e.g. for 5% damping, take $\xi = 5$). Similar, although slightly different, values are provided in IBC.

At this stage, the calculated deflection needs to be compared with the trial deflection of step (1), and the process iterated if necessary with revised values of spring stiffness and damping.

(5) The superstructure forces can then be calculated assuming a constant acceleration up the height of the building, which follows from assuming that the superstructure responds as a rigid block. Following Eurocode 8, this leads to

$$f_j = \eta m_j S_e(T)/(q = 1.5)$$

where f_j is the force to be applied at level j and m_j is the corresponding mass. Note that in this case, a structural reduction or 'behaviour' factor of $q = 1.5$ has been assumed; no special seismic detailing is then required. In IBC, the equivalent R value is taken as between 1 and 2, depending on the inherent ductility of the superstructure.

(6) The foundations should be designed for a shear force $F = \eta M S_e(T)$ at the level of the isolation system. In both Eurocode 8 and IBC, no structural reduction factor is generally allowed; this is to protect the foundations against the possibility of yield or damage in the design earthquake.

(7) The isolation system must be checked for a deflection greater than that corresponding to the ultimate limit state (ULS). In IBC, the design deflection from step (4) must be increased by a factor of 1.5 when checking the capacity of the isolation system. In Eurocode 8, the recommended increase factor for buildings is only 1.2, a significantly lower value, although it rises to 1.5 for bridges. In the authors' opinion, using an increase factor of less than 1.5 should be done with caution. This particularly applies if the site may be near the source of a large earthquake which can give rise to velocity pulses causing very large displacements.

(8) In general, the analysis should be carried out for both principal directions of the building.

Eurocode 8 Part 1 (CEN 2004) gives advice on when this simple type of analysis will suffice. Usually more sophisticated checks (discussed below) will be needed before finalising the design. However, in almost all cases, the very simple checks should be performed as an essential first stage.

13.5.3 More rigorous analysis

If the centre of mass of the superstructure does not coincide with the centre of stiffness of the isolation system, the resulting rotation about a vertical axis will increase the deflection of some of the bearings and this must be allowed for, for example by a simple 3-D model. Sensitivity analyses may be required to investigate the effect of variations in the stiffness and damping properties of the isolation system; Eurocode 8 recommends this should be done where the system properties may vary by more than $\pm 15\%$ from their mean values. The approximation of using an equivalent linear spring-damper system may also need to be checked; this can be done by carrying out a non-linear time-history analysis, while retaining the assumption that the superstructure remains rigid, which greatly simplifies the problem. Similarly, the effect of foundation flexibility on response can be checked quite easily if the 'rigid superstructure' assumption is retained.

However, assuming rigid response may lead to significant inaccuracy in a tall building, particularly where the level of damping in the isolation system is high. In these circumstances, the flexible modes of the superstructure may become coupled with the rigid body type of response, and the distribution of lateral forces will be changed from the simple, uniform distribution assumed in step (5) of the previous section. A response spectrum analysis, accounting for the superstructure modes as well as its rigid body response, may not capture this if there are high levels of damping in the isolation system. This is because the assumption that the superstructure modes are uncoupled becomes invalid. In this case, a time-history analysis is necessary using a complete model of the superstructure.

More detailed advice on the analysis of complex seismically isolated systems is given by Naeim and Kelly (1999) and Skinner *et al.* (1993). Mayes and Naeim (2001) provide a useful checklist of the analysis requirements of IBC for isolated buildings.

13.5.4   Ductility and seismically isolated buildings

The procedures discussed so far assume that in an isolated building, the foundations (everything below the isolation level) have a high degree of protection against yielding by designing them for unreduced elastic forces, while the superstructure remains essentially elastic, with limited reduction factors on elastic forces allowing mainly for overstrength rather than ductility.

These procedures clearly result in a building performance during the design earthquake which is superior to that implied by code design for fixed-based buildings, where considerable excursions into the plastic region are allowed during the ULS earthquake, implying commensurate damage. The possibility therefore exists of designing both foundations and superstructure in an isolated building for reduced lateral strength, and accepting that some damage will occur during the design earthquake. Eurocode 8 recognises this situation as 'partial isolation', although no further advice is provided. It is not referred to in IBC. Partial isolation might be particularly attractive when retrofitting an existing building by introducing seismic isolation, particularly where increasing the strength of the existing structural members would be difficult or expensive, but some ductility already exists.

In fact, the option of partial isolation of the superstructure is less straightforward than it seems. The isolation system works by considerably lengthening the period of the motions to which the superstructure is subjected. In these circumstances, very much larger plastic deformations are required in the superstructure to reduce the accelerations in the superstructure to the same extent as in the equivalent fixed-base structure. The effect is the same as that shown in Fig. 3.18, which shows that ductility is ineffective in reducing response in very stiff structures. Effectively, the base isolation acts to make the superstructure relatively stiff compared to the motions to which it is subject; this reduces the elastic response, but also limits the capacity for a ductile response to reduce internal forces. Ochiuzzi *et al.* (1994) provide further discussion.

As a result, a simple equivalent linear elastic analysis, using q or R factors appropriate for fixed-base buildings is not a possible option for designing partially isolated buildings. A time-history analysis, explicitly accounting for the non-linear behaviour of isolation system and superstructure is recommended, enabling the local curvature ductility demands in the superstructure to be quantified, and checked as sustainable for the level of seismic detailing provided. Since ductility demands may increase rapidly with increasing ground motion intensity after the superstructure starts to yield, a check on a 'maximum considered' as well as a ULS earthquake should also be performed.

13.6   Testing of bearing systems

Bearings for seismically isolated buildings are usually manufactured specially for a particular project, rather than being produced as standard off-the-shelf items. Therefore they need to be tested to confirm that achieved values of stiffness, damping and deformation capacity accord with design assumptions. One effect that may need investigation is 'scragging', a reversible phenomenon in rubber bearings whereby their initial stiffness reduces after a few cycles of loading but may subsequently recover after a period of time. Another possible effect to investigate is the influence of cycling rate on properties.

In Europe, the forthcoming EN 15129:200X (Anti-seismic devices) will provide product standards for seismic isolation bearings, including testing standards. In the USA, ASCE Standard SEI/ASCE 7-02 (ASCE 2002) provides testing requirements for seismic bearings in buildings. Tests for bridge bearings were referred to in subsection 13.2.3. Typically, testing is required for two prototypes of each type and size of bearing.

13.7   Active and semi-active systems

Conventional passive isolation systems have their limitations; their optimal characteristics depend on the precise nature of the seismic events to which they may be subjected, but the very nature of these events is that they are highly unpredictable. Over the past 20 years, work has been done on 'active' systems that can modify their structural characteristics in real time during an earthquake depending on the input motions and structural response actually experienced, in order to optimise response. Actively controlled structures have the potential to be more efficient and respond more tolerantly to a much wider range of ground motion types than is possible for passively protected structures. Clearly, however, there are potential pitfalls in the use of active systems to resist seismic forces. The control system must work very reliably under the chaotic conditions of an earthquake, which might occur many years after the construction of the building. Moreover, an active system by its nature requires a power supply, and this must also be available on demand during the few seconds of an earthquake. However far-fetched and impracticable the concept may seem, a lot of work has been done to develop it and a number of demonstration buildings with active control exist in Japan and elsewhere in South-East Asia; Nishitani (2000) lists 30 Japanese buildings with active control. Proponents of the system point out that modern aircraft rely on an automated, active control of the wind flaps to achieve aerodynamic stability, and this does not deter people from air travel, although of course aircraft systems are tested on a regular basis. Active systems have some way to go, however, before achieving the level of acceptance won by passively controlled systems such as seismic isolation or supplemental dampers.

Key elements of an active system are a controllable feature (or features) of the structure, a means of measuring ground motion input and structural response, and a controller with a feedback loop between the two which tries to optimise the setting of the controllable feature. One type of active control system is the active mass driver. The controllable feature here is a set of hydraulic actuators reacting against the inertia of a mass on wheels. The control system attempts to achieve forces in the actuators which act to reduce the motion induced by an earthquake. Kobori *et al.* (1991) describe the use of this system on a full-scale building in Japan. The system works well to control response to minor to moderate earthquakes and to wind-induced vibrations, but may not be sufficiently strong or powerful to resist a major earthquake, particularly one giving rise to a strong single pulse, as may occur near its source. The building described by Kobori *et al.* in fact relies on its conventional passive structure to resist the rare design event.

Semi-active control is a promising way of combining the advantages of active control while overcoming the drawbacks of reliance on a power source and the correct functioning of the control system. The controllable feature here is a set of variable dampers linking diagonal to horizontal bracing members (Fig. 13.10). The variability in damping can be achieved through viscous dampers with a variable size of aperture for the fluid flow, or alternatively friction dampers with a variable clamping force. At low levels of damping, the diagonal bracing is less effective and so the lateral stiffness and frequencies of the building decrease. Hence two dynamic characteristics – damping and period – are controllable, as are their distribution within the structure, which provides good opportunities for optimisation of response. The power requirements are much less than for the active mass driver; they involve opening or closing valves (or adjusting clamping

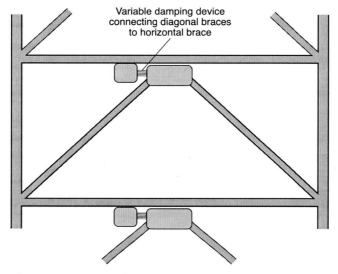

Fig. 13.10 Schematic diagram of semi-active variable damping control

forces) rather than needing to change the response of the building itself. With loss of power, the dampers can be set to a fail-safe position (such as fully engaged), so some of the reliability concerns with fully active control are met. However, the control system still has to work reliably.

Three World Conferences have been held on structural control which covers active and semi-active control against seismic effects; the most recent was held in Como, Italy in 2002 (Casciati 2003). The proceedings provide further information on the current state of the art.

13.8   References

AASHTO (2000). *Guide Specifications for Seismic Isolation Design*. American Association of State Highway and Transportation Officials, Washington DC.

ASCE (2002). ASCE 7-02. *Minimum design loads for buildings and other structures*. American Society of Civil Engineers, Reston, VA.

Beck J. & Skinner R. (1974). The seismic response of a reinforced concrete bridge pier designed to step. *Earthquake Engineering and Structural Dynamics*, Vol. 2, pp. 343–358.

Boardman P., Wood B. & Carr A. (1983). Union House – a cross braced structure with energy dissipators. *Bulletin of the New Zealand Society for Earthquake Engineering*, Vol. 16, No. 2, pp. 83–97.

Casciati F. (ed.) (2003). *Proceedings of the 3rd World Conference on Structural Control*. Wiley, Chichester.

CEN (2004). EN1998-1: 2004. *Design of structures for earthquake resistance*. Part 1: General rules, seismic actions and rules for buildings. Part 2: Bridges European Committee for Standardisation, Brussels.

HITEC (1996). *Guidelines for the Testing of Seismic Isolation and Energy Dissipation Devices*. CERF Report HITEC 96-02. Highway Innovative Technology Evaluation Center, Washington DC.

ICC (2003). IBC: 2003. *International Building Code*. International Code Council, Falls Church, VA.

Kobori T., Koshika N., Yamada K. & Ikeda Y. (1991). Seismic response control structure with active mass driver. Part 1: Design. Part 2: Verification. *Earthquake Engineering and Structural Dynamics*, Vol. 20, pp. 133–166.

Lang R., Kanada M., Hikone S. & Mole A. (2001). Stepping column system – a pagoda for the 21st century. In: Anumba C., Egbu C. & Thorpe A. (eds) *Perspectives on Innovation in Architecture, Engineering and Construction*. CICE. Proceedings of Conference at Loughborough University.

Mayes R. & Naeim F. (2001). Design of structures with seismic isolation. In: Naeim F. (ed.) *The Seismic Design Handbook*. Kluwer Academic, Boston.

Muir Wood R. (1988). Robert Mallet and John Milne – earthquakes incorporated in Victorian Britain. *Earthquake Engineering and Structural Dynamics*, Vol. 17, No. 1, pp. 107–142.

Naeim F. & Kelly J. (1999). *Design of Seismic Isolated Structures: from Theory to Practice*. Wiley, New York.

Nishitani A. (2000). Structural control implementations in Japan. *Proceedings of the 12th World Conference on Earthquake Engineering, New Zealand Society for Earthquake Engineering, Upper Hutt, New Zealand*, 30 January to 4 February.

Ochiuzzi A., Veneziano D. & Van Dyck J. (1994). Seismic design of base isolated structures. In: Savidis S. (ed.) *Earthquake Resistant Construction and Design*, Vol. 2, pp. 541–548. A. A. Balkema, Rotterdam.

Sharpe R. & Skinner R. (1983). The seismic design of an industrial chimney with rocking base. *Bulletin of the New Zealand National Society for Earthquake Engineering*, Vol. 16, No. 2, pp. 98–106.

Skinner R. I., Robinson W. & McVerry G. (1993). *Introduction to Seismic Isolation.* Wiley, Chichester.

Smith D. G. E. (1996). Base isolated structures. In: Blakeborough A., Merriman P. & Williams M. (eds) *The Northridge California Earthquake of 1994: a field report by EEFIT.* Institution of Structural Engineers, London.

14 Assessment and strengthening of existing buildings

'As the general awareness of earthquake risk increases and standards of protection for new buildings become higher, the safety of older, less earthquake-resistant buildings becomes an increasingly important concern.'

Andrew Coburn and Robin Spence.
In: *Earthquake Protection*. Wiley, Chichester, 2002

This chapter covers the following topics.

- Design strategies for strengthening
- Assessing the seismic adequacy of existing buildings
- Analysis methods for existing buildings
- Methods of strengthening
- Assessing earthquake-damaged buildings
- Special considerations for historic buildings
- Evaluating the seismic performance of large groups of buildings

14.1 Introduction

Even if all future new construction in seismic regions were built to conform to the best current standards (sadly, an unrealistic expectation), the existing stock of substandard construction would continue to pose a large risk for many decades. In mature economies, the rate of new construction is typically only around 1%, and although a higher rate applies in developing countries, the existing housing stock does not diminish very rapidly (Coburn and Spence 2002). It would therefore take many years before all the substandard construction were replaced. In fact, the risk they pose may increase for a number of reasons, including structural deterioration due to poor maintenance, weakening due to removal of internal partitions to create larger room sizes, and further weakening if earthquakes occur and the resulting damage is not repaired.

The existing housing stock in seismic regions is particularly prone to seismic defects. Houses have to support their gravity loads all the time, and the most severe wind load experienced in a ten-year period usually does not vary significantly from one decade to the next. Therefore, fundamental weaknesses generally become evident and local construction practices tend to adjust accordingly, and become (by and large) adequate for gravity and wind loads. A damaging earthquake, however, may occur less than once in a generation. Therefore, as housing styles and requirements change over time, inappropriate practices may develop which leave the construction vulnerable to earthquake damage. A classic example comes from the Turkish region to the east of Istanbul, which experienced extremely rapid economic expansion in the last decades of the twentieth century. The region was known to be highly seismic, but the seismic construction regulations applying to the area were not enforced; the desire for immediate accommodation was much stronger than any consideration of the threat from possible future earthquakes. Much of the construction was in fact highly unsuitable for earthquake country, and was decimated by the two major earthquakes which affected the region in 1999, resulting in great loss of life, destruction of property and resulting human misery.

A major earthquake often provides an impetus for strengthening substandard buildings in the surrounding area, but in many respects of course this comes too late. In a region assessed as seismic but which has not had an earthquake for many years, it is much harder to persuade building owners of the need for strengthening, particularly if they are non-resident landlords. Strengthening costs are usually a significant proportion of the initial construction cost; Coburn and Spence (2002) quote a range of 5–40% of the rebuilding cost. Moreover, the process is disruptive, often requiring temporary evacuation of the building. Nevertheless, extensive strengthening programmes have been conducted in many parts of the world, including California, Turkey, New Zealand and Japan. In some places, strengthening is enforced by statute, but this does not yet apply anywhere in Europe and more general statutory enforcement is regarded as necessary for seismic risk reduction. The publication of Eurocode 8 Part 3 (CEN 2005) may provide an opportunity for this to occur.

Removing the threat posed by substandard construction in seismic areas poses many complex social, financial and legal problems, as well as technical ones. The rest of this chapter concentrates on the engineering issues; a discussion of the wider issues is provided by Coburn and Spence (2002) and Spence et al. (2002).

14.2 Performance of strengthened buildings in earthquakes

Reports of the performance of strengthened buildings in strong earthquakes are limited. There are a number of reports that buildings retrofitted with concrete shear walls have performed well. One such building was noted in the 1985 Mexican earthquake (EEFIT 1986; Fig. 14.1). Another is Adapazari City Hall in Turkey, which was damaged by an earthquake in 1967. The subsequent retrofit programme involved addition of new concrete shear walls, and strengthening of existing beams, columns and shear walls. Adapazari was devastated by the 1999 Kocaeli earthquake, with over 40% of buildings damaged or destroyed. The City Hall however performed well; while there was some damage, including cracking of some infill walls and several broken windows, the building remained functional and was heavily used for recovery activities after the earthquake (EERI 2000). Over the past 20 years, a large number of buildings in highly seismic areas of Turkey, both damaged and undamaged, have been strengthened by the addition

Fig. 14.1 Building in Mexico City after the 1985 earthquake. The building had previously been strengthened by the addition of external shear walls

of shear walls, and it is likely that much more performance data will emerge as strong earthquakes affect these regions in future.

The performance of unreinforced masonry buildings in the Whittier (California) earthquake of 1987 is reported by Deppe (1988). These buildings had been part of an extensive strengthening programme carried out before the earthquake. Losses in the strengthened buildings were considerably reduced, compared to unstrengthened ones; however, they still suffered significant damage.

There are even fewer data on the performance of strengthened historic buildings in earthquakes. Feilden (1982) warns that 'a blind use of reinforced concrete [to strengthen historic buildings] can be disastrous'. Instances were reported after the 1997 earthquake in Umbria-Marche, Italy, of stiff reinforced concrete frames added to strengthen flexible masonry buildings, where the masonry subsequently shook loose, leaving the strengthening elements as the only survivors. However, other similarly strengthened masonry buildings in the same region performed well, and Booth and Vasavada (2001) noted that strengthening of some historic masonry buildings with concrete diaphragms greatly improved performance in the 2001 Gujarat earthquake. In the 1997 Umbria-Marche earthquake in Italy, damage to the Basilica of St Francis in Assisi was initially attributed by some commentators to the addition in the 1960s of concrete strengthening beams in the roof. Subsequent detailed investigations by Croci (1999) showed that in fact these played no part in the partial collapse of the Basilica roof that occurred during the earthquake.

14.3   Design strategies for strengthening

14.3.1   Performance targets for strengthening

There are a wide variety of circumstances under which strengthening may be considered for a building. It may have become unserviceable or unsafe due to earthquake damage. Alternatively, there may be no existing damage, but the owner has determined that the building poses an unacceptable threat to life in possible future earthquakes, or that strengthening is a worthwhile investment to protect against future financial losses, both directly from damage to the building fabric and indirectly from business interruption. The strengthening may be a stopgap measure to last a few years until the building is replaced, or it may be intended to preserve a historic building for many future generations. Strengthening may be forced upon the owner by statute, under a municipal programme for improving the seismic safety of a city's building stock.

It is clear that the design objectives will vary greatly, depending upon which circumstances apply. Both the required performance of the building under the design earthquake, and the probability of occurrence of that design event may vary widely. Eurocode 8 Part 3 (CEN 2005) recognises three performance targets for strengthening, as shown in Table 14.1. In US practice, FEMA 356 (FEMA 2000a) defines four performance levels, shown in Table 14.2, which are supplemented by more detailed descriptions of the associated damage levels in various types of elements, such as steel or concrete frames, masonry walls and non-structural components, such as glazing, lifts and computer systems.

Table 14.2 shows that the 'life safety' level defined by FEMA 356 is broadly equivalent to, but somewhat more damaging than, the performance level intended to be achieved in the event of a design earthquake by new buildings designed to the NEHRP provisions (BSSC 1997). The NEHRP provisions are a model US standard, which were the basis of the IBC 2003 code (ICC 2003). Similarly, the limit state of significant damage defined in Eurocode 8 is noted as being roughly equivalent to the performance required for new buildings under the design, 475-year return earthquake. Thus, if an existing building were brought up to the Eurocode 8 'significant damage' performance level for a 475-year return earthquake, it would broadly conform to the seismic standards for a new building. However, as noted previously, that performance level might not be appropriate, and adjustments could be made to either the return period of the earthquake considered for design, or to the performance level, or to both. Thus, if preservation of life were the main concern, the 'near collapse' limit state in Eurocode 8 might be the governing consideration, while 'damage limitation' might apply to cases where economic considerations were foremost. In either case, an appropriate return period for the design earthquake would have to be chosen, based on the annual and lifetime risk level to be achieved. Eurocode 8 Part 3 gives no guidance on the appropriate level of return period to associate with each limit state, although

Table 14.1 Performance requirements from Eurocode 8 Part 3 (CEN 2005)

Performance requirement	Description
Limit State of Near Collapse (NC)	The structure is heavily damaged, with low residual lateral strength and stiffness, although vertical elements are still capable of sustaining vertical loads. Most non-structural components have collapsed. Large permanent drifts are present. The structure is near collapse and would probably not survive another earthquake, even of moderate intensity.
Limit State of Significant Damage (SD)	The structure is significantly damaged, with some residual lateral strength and stiffness, and vertical elements are capable of sustaining vertical loads. Non-structural components are damaged, although partitions and infills have not failed out-of-plane. Moderate permanent drifts are present. The structure can sustain after-shocks of moderate intensity. The structure is likely to be uneconomic to repair.
Limit State of Damage Limitation (DL)	The structure is only lightly damaged, with structural elements prevented from significant yielding and retaining their strength and stiffness properties. Non-structural components, such as partitions and infills, may show distributed cracking, but the damage could be economically repaired. Permanent drifts are negligible. The structure does not need any repair measures.

the National Annexes to the code of a particular country may provide such guidance for use in that country.

14.3.2 Cost–benefit analysis of seismic strengthening

Coburn and Spence (2002) discuss the application of cost–benefit analysis to the choice of performance level. Achieving higher performance levels implies higher initial costs, because a greater level of upgrading is involved, but it also implies a potential for future savings due to reduced damage in future earthquakes. It also implies lower casualty levels. In fact, it is usually difficult to justify seismic strengthening on purely economic grounds by a cost–benefit analysis, even in highly seismic areas, and Coburn and Spence recommend an estimation of the cost per life saved as a useful additional tool in setting design performance levels. They also suggest that, when planning upgrading of a large collection of buildings (a city centre, for example) it is most cost-effective to target intervention to the most vulnerable buildings. Smith (2003) points out that cost–benefit analysis as the basis for strengthening may be misleading if based on the average level of loss, because of the skewed distribution of earthquake ground motions (see section 2.3). Thus, a low (or zero) level of intervention might give rise to the highest expected financial return, but leave a small but significant risk of a very large loss, if a rare event occurs. Instead, Smith recommends setting a level of unacceptable loss (which could be in financial terms or defined by numbers of lives lost) and then setting a design return period at which this risk must be reduced to an acceptable level.

Table 14.2 Damage control and building performance levels from FEMA 356 (FEMA 2000a)

	Collapse Prevention (CP) level	Life Safety (LS) level	Immediate Occupancy (IO) level	Operational level
	Severe	Moderate	Light	Very light
Overall damage				
General	Little residual stiffness and strength, but load-bearing walls function. Some exits blocked. Infills and unbraced parapets failed or at incipient failure. Building is near collapse.	Some residual strength and stiffness left in all storeys. Gravity load-bearing elements function. No out-of-plane failure of walls or tipping of parapets. Some permanent drift. Damage to partitions. Building may be beyond economical repair.	No permanent drift. Structure substantially retains original strength and stiffness. Minor cracking of façades, partitions and ceilings as well as structural elements. Elevators can be restarted. Fire protection operable.	No permanent drift. Structure substantially retains original strength and stiffness. Minor cracking of façades, partitions and ceilings as well as structural elements. All systems important to normal operation are functional.
Non-structural components	Extensive damage.	Falling hazards mitigated but many architectural, mechanical and electrical systems are damaged.	Equipment and contents are generally secure, but may not operate due to mechanical failure or lack of utilities.	Negligible damage occurs. Power and other utilities are available, possibly from standby sources.
Comparison with performance intended for (new) buildings under the NEHRP provisions, for the design earthquake	Significantly more damage and greater risk.	Somewhat more damage and slightly higher risk.	Less damage and lower risk.	Much less damage and lower risk.

14.3.3 Strengthening earthquake-damaged buildings

Buildings damaged in an earthquake require a somewhat different approach from undamaged ones, and in some ways the choices are clearer. Some form of intervention is needed, both to achieve an acceptable level of safety and to restore public confidence in the building. At the most basic level, a decision must be taken on whether to demolish and rebuild, or to strengthen. The relative cost of these two options, which depends on the degree of damage and the original deficiencies in the structure, is only one factor; the architectural merits (or demerits) of the building involved and the construction time and disruption associated with each option are also important.

Strengthening may involve merely reinstating the building to its pre-earthquake condition. However, usually an improved standard needs to be achieved. Often, a damaged building will have been evacuated, which makes radical structural intervention easier than for an undamaged, occupied building whose occupants are probably concerned with more immediate, every-day concerns than protecting themselves against a future, hypothetical earthquake. Cosmetic strengthening of a seriously damaged building, which merely hides the evidence of damage without addressing its causes, is of course an option to be avoided, but one which too often has been shown to have been associated with losses in subsequent earthquakes.

14.4 Surveying the seismic adequacy of existing buildings

14.4.1 Undamaged buildings

As a first essential step, a thorough survey is needed of a building where strengthening is being considered. Guidance on general methods of survey is given in a number of texts, for example Beckmann and Bowles (2004) and the Institution of Structural Engineers (1996). In the first instance, sophisticated equipment is unlikely to be required; a typical checklist is given in Table 14.3. A key aspect is often the standard of workmanship and the degree of corrosion

Table 14.3 Typical checklist of equipment for an initial building survey

- Notebook
- Pencils
- Camera (preferably digital)
- 300 mm spirit level
- Binoculars
- 3 m tape
- 20 m tape
- Compass
- Penknife
- Flashlight
- Face mask (for entering dusty spaces)
- A copy of the checklists from ASCE/SEI 31-3 (ASCE 2003)
- Ordinary hammer (for tapping surfaces for soundness, and if permitted, knocking off finishes, etc.)
- Schmidt hammer (for concrete buildings)
- Covermeter (for concrete buildings)
- Crack width gauge (for concrete or masonry buildings)

and other deterioration in structural elements. Beautifully finished buildings have been known to hide serious defects behind elaborate plaster work, which the owner may not be very keen for an engineer to chip away. The loft space and service areas are good places to start, since the unadorned structure can usually be seen there.

Often, the first survey needs to establish if there are any significant areas of seismic weakness. The checklists given in ASCE/SEI 31-3 (ASCE 2003) are very useful in this respect. In this document, two dozen types of building are identified, classified by their building material and structural form. For each type, a checklist of desirable seismic attributes is given; for example, in concrete moment frame buildings, the list includes the presence of a strong column/weak beam system, and the absence of flat slabs forming part of the lateral load-resisting system. Where the attribute is definitely absent, or cannot be confirmed, a 'concern' is raised, which needs to be further addressed. This could be done by detailed analysis or further testing and inspection. In some instances (e.g. the presence of short captive columns created by partial height masonry infill (Fig. 1.17)), it may be possible to rectify the concern directly (in the example, by separating the infill from the frame, or building it full height). Checklists are provided not just for the superstructure, but also for foundations and for non-structural elements. They are based on damage observed to typical US buildings, but have been used successfully in many other parts of the world.

More detailed investigations may include soil testing and testing of construction material strength. FEMA (2000a) gives advice on the available methods, and further guidance is given in Beckmann and Bowles (2004) and the Institution of Structural Engineers (1996). In the absence of construction drawings, detailed dimensional surveys will be required in order to perform an analysis of the structure; even if there are drawings, their accuracy needs to be confirmed by at least spot-checks on site. While overall dimensions are relatively easy to capture, details of structural elements are more problematic. In particular, the reinforcement in concrete members may be very difficult to establish and, in crucial areas like beam–column joints, sometimes impossible. A limited amount can be achieved by removing finishes and false ceilings, chipping away concrete cover, using a covermeter and so on, but the confidence with which the properties of the existing structure can be established will be important in the choice of strengthening strategy. The best that can be done may be to infer the likely details from a knowledge of construction practice of the time. Where there is uncertainty about the existing structure, most or all of the seismic resistance will need to be provided by new elements.

14.4.2 Surveying earthquake-damaged buildings

After a damaging earthquake, an urgent need exists to establish which buildings continue to be safe to use, and which should be evacuated. The Applied Technology Council (ATC) in California has published the ATC-20 series of documents containing guidance for rapid and detailed evaluation of earthquake-damaged buildings (of all types) to determine if they can be safety occupied. Included are the basic procedures manuals (ATC-20 and ATC-20-2), a field manual (ATC-20-1), a manual containing case studies of rapid evaluation (ATC-20-3), a training slide set (ATC-20-T), and a TechBrief concerning earthquake aftershocks and building safety evaluation (ATC-TB-2). Additionally, documents prepared under the ATC-43 project provide guidance on in-depth engineering evaluation and repair of earthquake-damaged masonry-wall buildings and concrete-wall buildings (FEMA 306, FEMA 307, FEMA 308), which are collectively available on the ATC-43 CD. Further details are provided on the ATC website (www.atcouncil.org/).

14.5   Analysis methods

14.5.1   Approximate initial methods

It is often useful to perform an initial crude analysis to get a rough idea of the likely need for strengthening. The 'Tier 1' analysis procedures given in section 3 of ASCE/SEI 31-3 (ASCE 2003) provide a first estimate of whether or not the strength, ductility and stiffness of a building are likely to be adequate. They are based on approximate equivalent static force procedures.

14.5.2   More detailed analysis

All the forms of analysis discussed in Chapter 3 – equivalent static, response spectrum, non-linear static (pushover) and non-linear time history – may at various times be appropriate for analysing existing buildings. Pushover analysis is particularly useful. This is because it allows the ductility demand to be quantified in the yielding regions of the structure, and checked against the available ductility actually provided. In designing a new building, the adequacy of the ductility supply is usually ensured by following standard detailing rules in codes of practice. Often, however, existing buildings have been designed to standards now considered inadequate, or they may have been substantially altered after construction in ways that reduced their seismic resistance. However, they may still possess some ductility, which (particularly when combined with strengthening measures) may prove adequate, although not complying with current code rules.

For this reason, both Eurocode 8 Part 3 (CEN 2005) and FEMA 356 (FEMA 2000a) provide extensive and complete guidance on non-linear static analysis, including data on the plastic rotation of hinges associated with the performance levels defined in Tables 14.1 and 14.2. These deformation data have been prepared for structures which do not conform to modern standards, as well as ones that do. In FEMA 356, the data are presented in convenient tabular form. In Eurocode 8 Part 3, they are in the form of equations for concrete, steel and composite construction, which are more versatile, but may be harder to use in practice; they are also acknowledged to be somewhat experimental in nature.

Eurocode 8 Part 3 provides for three basic types of analysis

(1) linear analysis, using equivalent static or response spectrum techniques
(2) non-linear analysis, using static (pushover) or non-linear dynamic time-history methods
(3) q-factor approach.

These three types of analysis are now discussed in turn.

(a) Linear analysis to Eurocode 8 Part 3

This type of analysis uses as input the elastic ground motion spectrum, unreduced by q factors. The forces obtained from the analysis therefore do not account for any yielding in ductile elements of the structure. Brittle elements, and brittle failure mechanisms such as shear in concrete, are checked according to the ratio $\rho = D/C$, where D is the force demand on the brittle element or mechanism from the linear elastic analysis, and C is the capacity of the element. Where ρ is less than 1, the element is clearly satisfactory, since the linear elastic analysis tends to overestimate forces by ignoring the beneficial effects of yield in the ductile elements. Brittle elements with ρ greater than 1 may still be satisfactory, but must be checked as capable of sustaining the plastic forces developed in the ductile elements, on capacity design principles.

Ductile elements are checked on the basis not of strength, but deformation. For flexural members, the curvature predicted by the analysis must not exceed the ultimate curvature of the member. Guidance on limiting values is given in the Annexes to Eurocode 8 Part 3, and (although not referred to by Eurocode 8) the tables in FEMA 356 referred to above may also be suitable. Similarly, the axial strain in axially loaded members can be compared with fracture strain.

Linear procedures take no account of the effects of yielding on the distribution of strains. In many cases, they may underestimate the plastic demands and therefore they are unsafe; this would for example be the case for a soft-storey structure, where most of the strains occurring after initial yield of the structure are concentrated in the soft storey in a way not predicted by linear analysis. For this reason, EC8 limits the use of linear procedures to cases where the ratio of maximum to minimum values of ρ over all 'ductile' primary elements of the structure does not exceed a recommended value of 2.5. In this calculation, only ductile members with $\rho > 1$ are taken into account. This is intended to ensure that ductile demand is reasonably evenly spread throughout the structure, limiting the possibility of serious underestimate.

(b) Non-linear analysis

In non-linear methods of analysis, whether static or dynamic, the effect of yielding on the distribution of forces and plastic strain demands is directly accounted for, and the limitations of the linear methods are avoided. The EC8 procedures for checking ductile and brittle members are however the same as for the linear analysis described above.

FEMA 356 (FEMA 2000a) treats non-linear analysis in a very similar way, and provides extensive advice on carrying out a non-linear static (pushover) analysis.

(c) 'q-factor' analysis

In EC8, a force-based analysis procedure is also allowed, although the guidance is somewhat limited. The procedure is the same as that used for new buildings; a global reduction factor (q factor) is applied to the results of an elastic analysis, and ductile elements and mechanisms must be checked for a resistance at least to the demands from this analysis. Brittle elements are then checked on a capacity design basis to show that they can develop the strength of the ductile elements.

The problem with such an analysis is that q factors are set on the basis that design and detailing conforms to a stringent set of code requirements. The design of many existing buildings for which strengthening is contemplated, however, differs markedly from current code rules; that is usually why the buildings need strengthening. EC8 provides that when checking the limit state of significant damage, a value of $q = 1.5$ and 2.0 for reinforced concrete and steel structures, respectively, may be adopted regardless of the structural type. The code allows that higher values of q may be adopted if suitably justified with reference to the local and global ductility which is available, evaluated in accordance with the relevant provisions of Eurocode 8 Part 1; no further advice is given. EC8 states that the q factor method is not generally suitable for addressing the limit state of near collapse, although it notes that it may be checked using q factors increased by 'about one-third' of those for the significant damage limit state. The limit state of damage limitation is addressed by checking storey drifts, in the same way as for new buildings.

FEMA 356 also provides a force-based method, and gives extensive information on the force reduction factors that apply to structures that are not code compliant. The reduction factors are applied on an element-by-element basis rather than a global basis; thus for example, a separate reduction factor would apply to a beam with inadequate confinement steel as opposed to a column in the same structure under high axial load. Separate reduction factors are given for each of the three performance levels in FEMA 356, namely Immediate Occupancy, Life Safety and Collapse Prevention (see Table 14.2).

14.6 Assessing element strengths and deformation capacities

The previous section discussed the analysis methods needed to assess the strength and deformation, or ductility, demands on the members of an existing structure. These then need to be compared with member capacities, to establish whether or not they are adequate.

The approach of both FEMA 356 and Eurocode 8 is similar. Member capacities are assessed on the basis of their *expected* values. In new buildings, lower bound or characteristic values are used, further reduced by partial material or capacity factors. Higher values are possible for existing structures because various factors, such inaccurate member dimensions or poor construction quality, can be directly measured and accounted for. However, other uncertainties apply, particularly where the original design documentation is not available, and both FEMA 356 and Eurocode 8 require that capacities are reduced by a knowledge factor, which reflects these uncertainties. For buildings where there is only limited knowledge of the original design, Eurocode 8 recommends dividing the expected capacity by 1.35, compared with 1.0 for cases of complete knowledge, and similar reductions apply in FEMA 356.

Various sources of guidance for establishing the expected capacity of existing buildings were described in section 14.4.

14.7   Methods of strengthening

14.7.1   General

Methods of strengthening are provided by various authors, including Coburn and Spence (2002), Feilden (1982) and Beckmann and Bowles (2004). FEMA 356 (FEMA 2000a) also provides extensive information on rehabilitation methods. Other references include ATC 40 (ATC 1996) and the ICBO guidelines (ICBO 2001). What follows are some introductory notes.

14.7.2   Some initial considerations

Initial considerations may include the following.

(a) Damaged or deteriorated elements are likely to need repair or replacement.

(b) Irregularities in plan leading to torsional response, caused for example by poorly arranged shear walls or asymmetric masonry infills, should be removed. Adding shear walls or cross-bracing to reduce the torsional eccentricity is an obvious possibility.

(c) Irregularities in elevation giving rise to soft or weak storeys should be eliminated by addition of suitable strengthening and stiffening elements.

(d) Lack of tying together of a building may well be a concern in masonry buildings, particularly at the connections between floors and walls. Local fixings between wall and floors or steel tie rods extending across the entire building may be options to address this.

(e) Non-structural elements need protection as well as the structure. Cladding elements, chimney stacks, parapets, light fittings, storage units and services may all need to be considered.

(f) Geotechnical aspects should be addressed, including the possibility of slope instability at or near the site, and foundation movements.

14.7.3   Strengthening options

Among the possible options are the following.

(a) Addition of shear walls

Additional concrete shear walls have been widely used to strengthen and stiffen inadequate reinforced concrete moment frame structures. The shear walls reduce the ductility demands in an earthquake on the beam and column frames, which then are much more likely to be able to continue supporting the gravity loads they have to carry. The shear walls also reduce the tendency of a weak or soft storey to form, and their stiffness provides protection to non-structural elements, particularly cladding. The method is particularly suitable for low-rise construction up to five storeys.

The concrete shear walls are typically formed within an existing concrete frame, with dowelled connections to the surrounding beams and columns (Fig. 14.2). Casting each lift of the wall so that it is well compacted up to the soffit of the beam above needs careful placing of the concrete, but is quite easily achieved in practice. Strengthening walls placed outside the frame are also possible, and overcome the compaction problems mentioned above, but are more architecturally intrusive, and it is harder

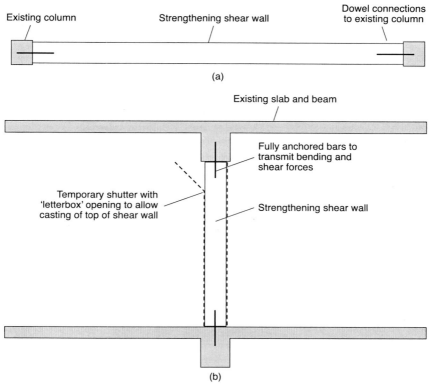

Fig. 14.2 Addition of concrete shear walls to a concrete frame building: (a) plan view; and (b) sectional elevation through strengthening shear wall

to achieve bonding to the existing structure. External buttressing walls have also been used, which may assist in keeping the building occupied while strengthening works are carried out, but the same architectural and structural difficulties apply.

Instead of reinforced concrete, the strengthening shear walls can take the form of infill masonry, which is particularly suitable if the strength shortfall is low, and the main objective is to remove eccentricities in plan or elevation. Recently, methods of strengthening existing blockwork panels have been developed, using carbon fibre reinforced bands fixed with epoxy mortar in X-shaped bands between the panel corners. These methods are similar in construction cost to additional concrete shear walls, but should prove much less disruptive to the continuing operation of the building.

The additional shear walls need to be provided with foundations that are stiff and strong enough to develop the required moments and shears at their bases, and this can give rise to requirements for substantial new footings, particularly where poor ground conditions occur.

(b) Cross-bracing
Adding steel cross-bracing to an inadequate concrete or steel moment frame building is an alternative technique to the addition of concrete shear walls. It

has a similar effect of relieving ductility demands on the existing frame, and protecting the non-structural elements. It may be possible to add an additional cross-braced steel frame relatively quickly to an existing building, minimising disruption.

The bracing members may be attached directly to the existing frame, which must then be capable of taking additional axial forces during an earthquake. This is the most efficient and least disruptive solution, but depends on the adequacy of the existing beams and columns. Alternatively, a complete new braced frame can be added, leaving the existing frame to carry only the gravity loads.

(c) Passive dampers

Cross-bracing can be added to existing moment-resisting frames via passive dampers, which serve to limit the additional forces that the existing frame has to take, while dissipating energy and so reducing response (Fig. 14.3). The dampers can take the form of viscous, hysteretic or frictional devices. Further information is given by Soong and Dargush (1997).

(d) Jacketing of concrete frame elements

Frames with inadequate confinement and shear strength can be surrounded by the elements in a confining jacket. The jacket may be made of steel plate, reinforced concrete or a composite material, such as polymers reinforced with carbon fibre, glass or kevlar. Annex A of Eurocode 8 Part 3 (CEN 2005) provides data quantifying the improvement obtained from these various jacketing methods.

It is usually possible to wrap the jacket around an entire column, producing an efficient confinement. The presence of a slab often makes this difficult or impossible for beams, and so jacketing is mainly useful for increasing shear strength, with little effect on the flexural ductility. Jacketing of beam–column joints is not possible, and this limits the applicability of the technique.

(e) Strengthening of steel moment-resisting frame structures

Since the weld failures at the beam–column joints of steel moment-resisting frames experienced in the Northridge and Kobe earthquakes, discussed in Chapter 9, there has been an extensive research effort to develop upgrading methods to avoid such failures in future. A favoured option is to prevent the formation of plastic hinges at the welded joints. This can be done by welding additional plates to the beams at the joints, so that the plastic hinge forms in the relatively weaker section of beam away from the joint. Alternatively, the same result can be achieved by cutting away the flanges of the beam some distance from the joint (Fig. 9.13), although the consequences of this weakening of the beam need to be checked. FEMA 351 (FEMA 2000b) gives extensive recommendations based on US research.

(f) Strengthening of floors

Floors play a vital role in seismic resistance by distributing the inertial forces generated in an earthquake back to the lateral resisting elements, and by tying the entire structure together. The strength and stiffness of timber floors can be improved by screwing additional plywood sheets to the floor joists and providing

(a)

(b)

Fig. 14.3 Hysteretic steel dampers added as retrofit: (a) detail of damper; and (b) general view (Perry et al. 1993)

additional blocking elements between joists. Concrete floors may be strengthened by the addition of a concrete screed. The additional gravity loads that the floor bears as a consequence must of course be checked.

(g) Reinforcing wall-to-floor connections
The connection of a timber floor to supporting masonry walls may be improved by the addition of a timber edge beam around the perimeter of the floor, screwed or bolted to the existing floor and attached to the walls with anchor bolts or through-bolts.

Fig. 14.4 Seismic isolator installed in the ground-floor column of an existing concrete frame structure. Seismic isolation was chosen for retrofitting this busy airport terminal in Turkey, since installation of the lead–rubber isolation bearings could continue while the terminal was still in use

(h) Guniting of masonry walls

Masonry walls can be strengthened adding a thin layer of mortar to one or both faces. The layer is strengthened with a light mesh reinforcement, and the mortar is usually applied at high pressure to improve compaction and bonding to the masonry, a technique known as guniting.

(i) Seismic isolation

As discussed in Chapter 13, this technique has the potential to protect both structure and non-structure from earthquake damage, while minimising the intervention required to a single isolation plane (Fig. 14.4). It is therefore particularly suitable if preservation of the existing architecture and (perhaps) minimising disruption to the continued occupation of a building are important.

14.8 Special considerations for strengthening earthquake-damaged buildings

14.8.1 Categories of damage

Rapid categorisation of buildings as safe or unsafe is needed after an earthquake; subsection 14.4.2 discussed the assessment techniques and tools available. In many cases, the most heavily damaged buildings need to be demolished, and retrofitting is only an option for lower grades of damage.

14.8.2   Methods of repair

The performance goal of the strengthening needs to be defined. Is it merely to reinstate the building to its condition before it was damaged by the earthquake, or is it to be upgraded and if so to what level?

Damaged members need to be reinstated at least to the level that they can safely carry their gravity loads; otherwise they must be replaced. Grouting under pressure of cracks in concrete and masonry with epoxy mortar is a well-established technique which is quite reliable in reinstating the concrete to its previous capacity. Jacketing and plating, as previously described, can also be used to reinstate and strengthen damaged elements. Additional elements can then be added to take the demand away from inadequate members under conditions of earthquake loading.

Where foundations have settled during an earthquake, it may be possible to reinstate by jacking up the structure at the points of settlement, but almost certainly this will need to be accompanied by underpinning or other remedial measures to the foundations, particularly if the settlement was caused by liquefaction.

Recommended procedures for upgrading steel moment-resisting frame buildings damaged by earthquakes are provided by FEMA 352 (FEMA 2000c). Advice on repair of earthquake-damaged concrete and masonry wall buildings is provided by FEMA 308 (FEMA 1999).

14.9   Upgrading of historic buildings

A number of special considerations apply to historic buildings, including the following.

(a) Their historical significance implies that they need preserving for many future generations, so the return period of the design needs to be correspondingly long.

(b) The upgrading needs to pay particular attention to the cultural and architectural values of the original construction. Balancing the need to prevent future damage while preserving the original delight and value of an old building is perhaps as much an art as a science, particularly where the 'original construction' has been built and altered over many years.

(c) Often, it will be difficult to gain full knowledge of the properties of the materials and methods of construction.

(d) As far as possible, strengthening measures should be reversible, so that they can be removed and modified without damage to the original, if in future more effective measures are developed. In practice, this may be difficult to achieve fully, but the strengthening measures should 'respect, as far as possible, the character and integrity of the original structure' (Feilden 1987).

Feilden (1982, 1987) has written two classic texts on the restoration of historic buildings. In the second of these two texts (Feilden 1987), appendices are included giving conclusions and recommendations prepared by ICCROM (International Centre for the Study and Restoration of Cultural Property, Rome). These contain much practical advice on the issues involved. A useful case study of the upgrading of a historically important building dating from 1914, using seismic isolation, is provided by Steiner and Elsesser (2004).

14.10   Assessment of large groups of buildings

The chapter so far has dealt with the assessment and strengthening of individual buildings. Rather different techniques apply when assessing large groups of buildings. Cases include assessment of the building stock of a city centre to assist in formulating a seismic strengthening policy, or assessment by an insurance company of the seismic risk associated with its portfolio of buildings. The average annual loss or the 'maximum credible loss' (an ill-defined term) may be of interest. Alternatively, an estimate may be needed of the loss due to a given earthquake 'scenario' – that is, an earthquake of a given magnitude at a specified distance.

The technique used is to divide the building stock of interest into a few categories with broadly similar characteristics, for example low-rise unreinforced masonry, three- to five-storey concrete moment frames and so on. An estimate is then made of the likely distribution of damage within each class of building for a given level of ground shaking. These damage estimates can be based empirically on observations of damage in past earthquakes; usually the ground motion is described in terms of intensity, using the Modified Mercalli Intensity (MMI) or some other scale. ATC 13 (ATC 1985; supplemented by Anagnos *et al.* 1995) gives one Californian source for such data. An alternative approach is provided by HAZUS (FEMA 2003), which describes the ground motion in terms of a response spectrum rather than by intensity, and uses a non-linear static (pushover) method of analysis. Further discussion of the two methods is provided by Booth *et al.* (2004).

14.11   References

Anagnos T., Rojahn C. & Kiremidjian A. (1995). *NCEER–ATC Joint Study on Fragility of Buildings.* Technical Report NCEER-95-0003. National Center for Earthquake Engineering Research, State University of New York, Buffalo, NY.

ASCE (2003). *ASCE/SEI 31-3: Seismic evaluation of existing buildings.* American Society of Civil Engineers, Reston, VA.

ATC (1985). *ATC 13: Earthquake damage evaluation data for California.* Applied Technology Council, Redwood City, CA.

ATC (1996). *ATC 40: Seismic evaluation and retrofit of concrete buildings.* Applied Technology Council, Redwood City, CA.

Beckmann P. & Bowles R. (2004). *Structural Aspects of Building Conservation* (2nd edn). Elsevier, Oxford.

Booth E. & Vasavada R. (2001). *Effect of the Bhuj, India earthquake of 26 January 2001 on heritage buildings.* Web publication www.booth-seismic.co.uk/#Gujarat.

Booth E., Bird J. & Spence R. (2004). Building vulnerability assessment using pushover methods – a Turkish case study. *Proceedings of an International Workshop on Performance-based Seismic Design*, Bled, Slovenia, 28 June to 1 July.

BSSC (1997). *NEHRP recommended provisions for seismic regulations for new buildings and other structures.* Buildings Seismic Safety Council, Washington DC.

CEN (2005). EN1998-1: 2005. *Design of structures for earthquake resistance.* Part 3: Assessment and retrofitting of buildings. European Committee for Standardisation, Brussels.

Coburn A. & Spence R. (2002). *Earthquake Protection.* Wiley, Chichester.

Croci G. (1999). Strengthening of the Basilica of St Francis at Assisi. Reported by Booth E. in *SECED Newsletter*, Vol. 13, No. 4, pp. 8–9. SECED, Institution of Civil Engineers, London.

Deppe K. (1988). Whittier Narrows, California earthquake of October 1, 1987: evaluation of strengthened and unstrengthened unreinforced masonry in Los Angeles city. *Earthquake Spectra*, Vol. 4, No. 1, February, pp. 157–180.

EEFIT (1986). *The Mexican Earthquake of 19 September 1985*. SECED, Institution of Civil Engineers, London.

EERI (2000). Kocaeli, Turkey, earthquake of August 17, 1999: reconnaissance report. *Earthquake Spectra*, Supplement A to Vol. 16. EERI, Oakland, CA.

Feilden B. (1982). *Conservation of Historic Buildings*. Architectural Press, Oxford.

Feilden B. (1987). Between two earthquakes: cultural property in seismic zones. *Proceedings of an International Conference on Conservation Rome (ICCROM)*, Rome.

FEMA (1999). FEMA 308. *Repair of earthquake damaged concrete and masonry wall buildings*. Federal Emergency Management Agency, Washington DC.

FEMA (2000a). FEMA 356. *Prestandard and commentary for the seismic rehabilitation of buildings*. Federal Emergency Management Agency, Washington DC.

FEMA (2000b). FEMA 351. *Recommended seismic evaluation and upgrade criteria for welded steel moment-frame buildings*. Federal Emergency Management Agency, Washington DC.

FEMA (2000c). FEMA 352. *Post earthquake evaluation and repair criteria for welded steel moment-frame buildings*. Federal Emergency Management Agency, Washington DC.

FEMA (2003). *HAZUS technical manual*. Federal Emergency Management Agency, Washington DC.

ICBO (2001). *Guidelines for seismic retrofit of existing buildings*. International Conference of Building Officials, Whittier, CA.

ICC (2003). IBC: 2003. *International Building Code*. International Code Council, Falls Church, VA.

Institution of Structural Engineers (1996). *Appraisal of existing structures*. IStructE, London.

Smith W. (2003). Criteria for strengthening buildings: cost–benefit analysis is misleading. *Bulletin of the New Zealand Society for Earthquake Engineering*, Vol. 36, No. 4, NZSEE, Wellington, NZ.

Soong T. & Dargush G. (1997). *Passive Energy Dissipation Systems in Structural Engineering*. Wiley, Chichester.

Spence R., Peterken O., Gülkan P., Booth E., Bommer J. & Aydinoğlu N. (2002). Earthquake risk mitigation: lessons from recent experience in Turkey. *Proceedings of the 12th European Conference on Earthquake Engineering*. Elsevier Science, Oxford.

Steiner F. & Elsesser E. (2004). *Oakland City Hall Repair and Upgrades*. Earthquake Engineering Research Institute, Oakland, CA.

Index

Page references in italics are to illustrations and diagrams. Place names refer to earthquakes cited in the text.

weak storeys *see* soft storeys
welds
   beam–column joints 201–202, *202*
   brittle failures at 184–185, *184*
   low-cycle fatigue 185, 190–191
   specifications and procedures
      191–192

wind
   motions, contol 112, *113*
   seismic isolation 115

yielding responses
   dealing with 62–63
   significant 62